PLANTS, AGRICULTURE,
AND HUMAN SOCIETY

PLANTS
AGRICULTURE
^{AND} HUMAN SOCIETY

William Norman Richardson

Thomas Stubbs

W. A. Benjamin, Inc.

Reading, Massachusetts ● Menlo Park, California
London ● Amsterdam ● Don Mills, Ontario ● Sydney

Drawings and cover illustration by Fran Milner.

Library of Congress Catalog Card No. 77-72644
ISBN 0-8053-8215-1
ABCDEFGHIJK-HA-798

W. A. Benjamin, Inc.
2727 Sand Hill Road
Menlo Park, California 94025

ABCDEFGHIJ-HA-798

For the Pritchards —
Peter, Sibille, Sebastian, and Dominic
and to the memory
of Dr. Edith Dorothea Witt

PREFACE

The development of civilization has depended on human interactions with plants. Even the most primitive of societies —with a few exceptions—have relied heavily on plants for food, housing materials, medications, clothing, and other uses. More advanced cultural evolution was made possible by the practice of systematic agriculture.

In *Plants, Agriculture, and Human Society* we explore many of the ways in which humans make use of plants. Agriculture is presented as a complex interaction between human beings and their ecosystem, each being dependent on the other. We believe this book will be meaningful as a text for botany courses which deal with the interactions of plants and humans. It will also be useful as a supplement in anthropology courses focusing on this approach.

Unit One begins with a brief description of botanical taxonomy and a coverage of the evolution of the Angiosperms, that class including most plants of agricultural importance.

In Unit Two we emphasize the origins of agriculture. All major human societies of the past and of the present have had an agricultural base. Early agriculture provided the comparative abundance of food that freed hunting and gathering societies from the need to be in continuous pursuit of food. The development of agriculture has made it possible for an increasing percentage of societies' members to expend their energies in other areas. Today a relatively small number of people in the developed nations are responsible for producing food. The remaining citizens may devote themselves to other occupations.

Unit Three covers material in the realm of traditional economic botany: important crops, forestry products, and other commercial uses of plants.

Unit Four is a collection of chapters that we have termed "ethnobotany." We discuss approaches to gathering data in the field, social dynamics associated with the distribution of agricultural products in nontechnological societies, the nutritional value of ethnic diets, and the use of plants containing mind-altering substances. The latter have figured in some way in most cultures; until recently there was usually a religious connotation.

In the final unit we examine the causes of and potential remedies for global food shortages. It is ironic that improved agricultural techniques have been partially responsible for aggravating the problem by permitting the increase of human populations already living marginal existences. When confronted with related facts and figures, the gravity of this continuing crisis is evident.

Further readings are suggested at the end of each chapter. For the most current perspectives, the student is encouraged to consult *Science, Scientific American,* and other periodicals.

It would be difficult to single out everyone who has been influential and helpful to the formation of this book. Particular thanks are due to Dr. Arthur Nelson and Dr. Van Norman for their conscientious readings and constructive advice as the manuscript advanced. We are indebted to the people at W. A. Benjamin for their dependable assistance and confidence while the book has been in production.

W. N. Richardson
T. Stubbs
October 1977

CONTENTS

INTRODUCTION

In the past, the interactions between agricultural plants and civilization have been considered from the point of view of human beings as biological engineers, manipulators, and consumers. The technological developments of recent decades have had an overwhelming impact on food production, food preservation, the genetic engineering of new plant crops, and the distribution of food. These advances have been so impressive and far-reaching that we now take them for granted. Because technology has accomplished so much, we have an attitude of quiet complacency and optimism—a belief that human beings are now indeed in full control of feeding the increasing world population, of maintaining current levels of social and cultural development, and of insuring our future growth and survival as a species.

With the beginning of the environmental consciousness of the 1970's, information accumulated increasingly and continuously that our position of technological and environmental omnipotence was not secure or safe. Starvation and hunger have not vanished. Over half the world's population still suffers from malnutrition. The technological masterpieces of the Green Revolution—such as hybrid corn, wheat, and rice strains—offer only a partial solution. Once again, like generations before us, we must view the perpetuation of our social, cultural and economic systems—the very survival of our own species—as directly and critically dependent on food production. We have yet to gain control of the life-support systems essential to the continued existence of humankind.

During the last thirty years, we have become increasingly dependent on the technological developments relating to agricultural production. Yet these developments brought new and disquieting problems. Over the last 15 years it has become more and more evident that the use of pesticides is at best a mixed blessing. Slowly the information has accumulated on the effects of the widespread use of DDT and other organochloride compounds. Today, we know that such substances spread throughout our biosphere—regardless of the site of application. Strains of insect pests not only become resistant to the effects of the poisons, but actually, in some instances, develop metabolic dependence on a pesticide. Also, the application of insecticides is a threat to the species diversity of major plant and animal groups—including phytoplankton, marine mammals, and many birds. For example, DDT decreases the primary productivity of some land crops and of the phytoplankton in the oceans. You carry in your own body fat a concentration of organochlorides of about 12 parts per million. If you lived in India, that concentration would be 26 parts per million because the use of pesticides there has been more extensive. What are the implications? Scientists know that concentrations of these chemicals in experimental animals result in behavioral and reproductive disorders, but we cannot predict the total effect on our own species. That story is locked in the future and in human evolution.

The genetic engineering of plant crops and the cultivation of endless "amber waves of grain" stretching from Ohio to the foothills of the Rockies once appeared to be the answer to the necessary increase in crop production to feed the growing world population. However, just as weeds love the homogeneous blue-grass lawn, so do insect pests and fungal diseases respond with enthusiasm to fields of genetically similar crops. When predators or disease organisms confront a population that is diverse in its response to their attack, some members of the population succumb; but many survive. For example, when the fungal corn blight occurred in the midwestern U. S. in 1971, great losses were sustained because of the genetic similarity of the U. S. hybrid corn crop. The biological axiom is: "In diversity lies survival." The axiom of agriculturalists and economists has been: "In homogeneity lies productivity." Therein lies the conflict and the dilemma.

Recently scientists have calculated that, agriculturally, we are living on borrowed time with the increasing use of mechanized agricultural practices throughout the world. These calculations are based on simple energy considerations. To come out ahead, one must get out of a system more than one puts in. Consider the input of the raw materials involved in modern agricultural production—materials such as steel, petroleum products, chemical fertilizers, and pesticides. We are actually getting less in the long run than we are investing in the apparently successful effort. Put simply, if the developed countries continue to rely on highly mechanized methods for food production, what happens when the fossil fuels are depleted? It has been demonstrated that more fossil fuel is consumed in calorie equivalents—in mechanized planting, plowing, fertilizing, and harvesting—than is yielded by the crops on which the energy is expended. A technological optimist might reply that energy alternatives will be found for petroleum products and that technological advances will continue to offer solutions. This, of course, is a hope and not a solution.

What of the other problems created by the current approach to agricultural production? Mechanized farming necessitates repeated plowing, fertilizing, irrigating, and spraying for pests. What of the topsoil that is continuously lost to the atmosphere in plowing, creating dust clouds that change weather patterns? What of the nitrates and phosphates that are needed in increasing quantities to meet the growth requirements of the genetically engineered seeds? When these chemicals find their way into either fresh- or saltwater bodies, the productivity of the plant and animal communities living there is altered.

In response to these questions, the technological optimist and the concerned ecologist both ask: "What can we do other than what we are presently doing?" There are mouths to feed. Indeed, the world population will probably double during the next 35 years—from 4 billion in 1975 to about 8 billion by the end of this century. Despite modern agricultural technology, despite the international monetary loans for development of mechanized agricultural irrigation projects and new chemical fertilizer plants in the developing nations (DN's), many of these new mouths will starve unless other changes can be brought about.

Consider the interactions of three critical variables in human life: population, environment, and nonrenewable natural resources. In our opinion, these variables probably cannot be transcended by any technology. What may help human beings to overcome the problem described, at least in part, is a realization of the limits of planet Earth and the inescapable finiteness of its human family. We must reorder values and priorities regarding our resources and our environment and promote some understanding that human beings always have been, are presently, and always will be totally dependent for a food supply on an equilibrium between humankind and the environment. It must be our goal to preserve and improve the quality of the environment for future generations. Short term goals must be critically examined to avoid irreparable losses for future generations.

We, the authors of this book, are not agronomists; nor are

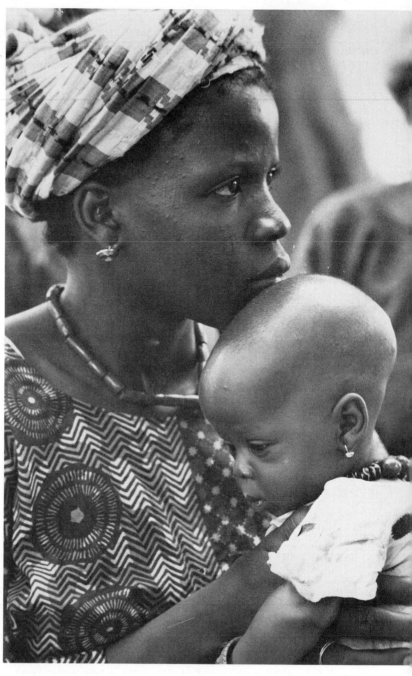

The dilemma of human beings' search for food is illustrated by this Nigerian mother and her baby awaiting consultation at a nutrition clinic. Diet deficiencies are due not only to food shortages, but also occur because people do not know the basics of sound nutritional practice. (FAO photo)

we technological optimists. We believe deeply that there are certain critical biological principles which have determined human beings' experiments with agriculture in the past. During the technological development of the last few decades, these principles were ignored or forgotten. Yet we believe that these same principles are applicable to present problems, and that the understanding and integration of these principles into a new approach to agriculture may well prove essential to human survival in the future. These principles are:

1. The interactions between plants and human beings over the last 100,000 years—and especially over the last 14,000 years since the transition into an agricultural lifestyle—are coevoluntionary. That is, *both* plants and human beings are changed by the interaction.
2. Agricultural production and stability have in the past and always will be, in our opinion, essential to the development, stability, and survival of human cultures.
3. There are biological limits to the extent that human beings can manipulate their environment. The understanding and respect of these limits will influence our biological and cultural survival.

In our current attitude toward agricultural production, we lack time-space perspective. Influenced by the Judeo-Christian perspective, we have viewed ourselves as apart—separate from the rest of the biological world. In the past we have cut ourselves off from our environment through our cultural activities. What we have overlooked is the fact that, through our actions toward the environment, we change not only the environment—we change ourselves. For there are continuous processes working—deeply rooted in the evolutionary mechanisms of all living things—by which the effector is changed as well as the effected. Such relationships are well known to physical anthropologists who document the human ascent from primate relatives. For example, diet affects tooth kind and pattern, and can even determine enzyme patterns among human beings in a given geographical area. As we will discuss, foods also determine and dictate cultural patterns.

In the Neolithic transition from hunting and gathering to a sedentary and increasingly agricultural way of life, more abundant food supplies permitted the diversification of human activity as many individuals in a society were freed from the necessity of food production. As individuals took up other occupations, this led to the increasingly complex development of human cultures.

In the developing countries of the world today, 60% to 80% of the population is still directly engaged in agriculture. In the technologically developed countries the percentage of the population involved with agricultural production is below 10% in many economies. It is a fact that efficient and surplus food production makes possible this diversity of culture. Without that excess production, industrialization and the spread of technology would slow down. There would be a shift, both in productivity and social priority.

After 10,000 years of development, agricultural production is still the base of civilization. The developing international food crisis will be experienced most immediately in the developing nations, where its impact will be most severe. However, in the developed countries trade patterns and priorities will also change. The focus will be on the basic commodity: food. At present there is a seemingly inevitable tidal wave of overpopulation in the world's future. If food production falls during the coming decades, international trade and politics will reflect the profound influence of shortages of commodities which we have taken for granted for far too long.

It is critical at this time that all branches of science, including the study of agriculture, adopt an *ecological perspective*. We must view our biosphere as a long-term life support system, and we must respect the natural limits of that system. In the developed countries, if we do not adopt this perspective we may exhaust our water supplies, lose priceless topsoil by mechanized practices, turn our lakes and rivers and estuaries into lifeless sewers, and poison ourselves as well as our insect pests. In all areas of agricultural development, a wider understanding of intrinsic relationships among components of living systems is needed. Before imposing stresses on the biosphere, both the long-term and short-term implications must be examined. We must realize that every stress imposed has an ultimate effect on our own health and long-term social welfare.

Human population growth must be limited in the developing nations (the developed countries having more or less achieved zero population growth). Until population growth is contained, we cannot put ourselves in harmony with the resources of the world system in which we live. Without a solution to the population problem, there are only short-term answers to the other concerns discussed above.

It is our belief that we are at a turning point in human cultural history. The present and future interrelationship between plants and humankind is critical to our biological and cultural welfare. Much of human history and culture still depends on the same grains that nurtured the earliest civilizations: corn, wheat, and rice. The world food crisis is forcing agriculturalists and biologists to explore alternative and often novel food sources, new methodologies, and new value systems. The problems of the developing nations are different in many ways from those of the developed nations. In this book both will be considered. However, it is clear that the developed nations must take the initiative in making others aware of the crisis facing us and in developing a technology to cope with the present dilemma—a technology with both long-term and short-term goals.

In this book we confront a variety of problems. Some, involving human beings and food, have not changed for millenia. Other problems which we discuss were totally unforeseen 10 years ago. Current technology (and most probably future technology as well) does not and cannot hold all the answers. But as long as there is a human being who is willing to plant a few grains of wheat, there is also hope.

The purpose of Unit One is to introduce readers to the scope of the text, and to acquaint them with some basic botanical taxonomy and principles. Chapter One explains our premise that a course about plant and human interaction should integrate related areas of study. Chapter Two is a summary and review of the structure of angiosperms, the large group of modern flowering plants which includes most of our agricultural stocks. Chapter Three describes the evolution and some co-evolutionary aspects of angiosperms. Although these chapters are relatively independent, they combine to provide background material to broaden the reader's understanding of basic botany, and to prepare the reader for more specific discussions in later chapters.

UNIT 1

CHAPTER

1

THE MERGER OF BOTANY, ANTHROPOLOGY, DEMOGRAPHY, AND ECONOMICS

A traditional approach to the study of plants and human beings focuses on the cultivation of various crops, the productivity of those crops, and the technological means for increasing that productivity. Gross yield, nutritional content, and enhancing the external appearance of certain fruits and vegetables were the main concerns. This book takes a somewhat broader approach—exploring relevant aspects of the relationships between human beings and the plants they use.

Differing Viewpoints   The Anthropologist's Perspective. From an anthropologist's point of view, the production, storage, and distribution of crops has always been, from earliest times, the foundation for the rise and endurance of extensive material cultures (the sum of cultures' material artifacts). Of course cultural groups may develop elaborate nonmaterial traditions. An example is the complex religious and social structure of Australian Aboriginal societies, which are currently and were in the past materially poor. However, the key element in expanding a society's material culture is *assured food supply*, and the presence of necessary raw materials. During most of the two million years human beings have been upon the earth, they lived in primitive bands as hunters and gatherers. Human potential was narrowly limited, for food-getting required the work of a majority of the members of society. Only when agricultural practices were developed and excess food supplies produced and stored, did human beings gain the free time to elaborate the material aspects of their cultures.

The Biologist's View. The biologist's perspective is concerned primarily with the interactions of ecosystems and organisms; or, the feedback among organisms and the environments they inhabit. A very basic biological principle is this: There is no such thing as a unidirectional effect. When two organisms interact with each other, *both* are affected. The same may be said for the relationship of any organism—including our species—with its environment. For example, the dentition of our ancestors changed as they adopted an omnivorous diet. In the Galápagos Islands off of Ecuador, Darwin's finches exhibit bill types suited to their food requirements. These interactions between organisms and their environment are a continuous process still going on today; not merely historical incidents seen in fossils, or static characteristics of current populations of plants and animals.
Human beings have effected extensive changes on the

plants they have used since our agricultural "experiment"*
began about 10,000 years ago. By artificial selection, albeit
often unconscious selection, human beings have brought
about significant changes in the plants that they cultivate—
changes in structure, productivity, nutritional content, and
the mechanisms of reproduction of the plants themselves.

Economists and Demographers. Economists and demographers have yet another perspective on agricultural botany. They study the subject in the context of global dilemmas.
Their focus is on the economics and demographic factors of
world food production (see Unit V). They ask: What is the
world food crisis? Why is there one? Will technology provide
satisfactory solutions?

Just as all natural systems are limited by the availability of
resources, so human (and other) populations are limited by
the quantity of available food. Available food is, in turn,
limited by available resources such as solar energy, water,
nutrients, photosynthetic efficiency, and the acreage of arable
land. Famines may occur even when the production of food is
adequate for a population, if there are no effective means for
food storage and distribution. In considering the world food
crisis, it is necessary to take these many related factors into
account, as well as the actual quantity of food being produced. In any nation, however industrialized and economically well-off that country may be, its success is based on
efficient agriculture—whether its own, or another country's.
Equally true is the assumption that no country can survive
indefinitely if it cannot meet the food requirements of its
people.

The Influence of Agriculture   In this book we will also deal with the influence
of agriculture in both the developed and the developing
countries. We will examine its effect upon the sustaining
capacity, the quality, and the diversity of the various ecosystems in these countries. Until recently, most agronomists
and technologists viewed increased production as the goal in
and of itself, and many still do. Unfortunately, increased
agricultural yield can no longer be treated in so one-
dimensional a way. We are presently confronted with the
prospect of decreasing environmental quality as a direct result of such attitudes. In the long run, the goal of increasingly

*We refer to the development of agriculture as an experiment because hu-
man interactions with plants are an evolving process, and a myriad of
problems in food production and agriculture are currently threatening to
overshadow former successes.

Above *Rice paddy in Sri Lanka (formerly Ceylon). More land is cleared for agriculture each year; yet there are intrinsic environmental limits to the amount of land that can be given to agriculture if an environmental balance is to be maintained. In many parts of the world, such a balance is not considered. (FAO photo)*

Left *Grain storage basket, Ivory Coast. (FAO photo)*

larger yields will inevitably reduce the amount of productive land. It is a self-defeating goal. The repeated plowing characteristic of mechanized agriculture, in which a field may be plowed up to 14 times between planting and harvesting, causes the loss of considerable quantities of precious topsoil, by destroying the soil structure and predisposing the land to erosion. In tropical areas, the destruction of jungle to create crop-producing land often causes *lateritic* soil (from the Latin *later*, meaning brick). Such soil exposed to the tropical sun soon becomes rock-hard and is useless for agricultural purposes (see Chapter 22).

The quality of both fresh- and saltwater may be altered by the enormous quantities of nitrates and phosphates that are required by genetically-engineered, high-yield crops. The presence of these chemicals in the run-off can affect the entire ecological chain—from water creatures to human beings. Even climatic changes occur as a result of land abuse through the cultivation of "marginal" lands, such as semi-desert, causing the land to become even starker than it was before agricultural attempts. The Sahara desert now encompasses areas where grain once grew in abundance to supply the Roman empire. These formerly productive areas became wasteland principally because of poor agricultural practices and deforestation which altered rainfall patterns. The "dust-bowl" of the United States is another example of an area that became agriculturally worthless during the 1930's as a result of destructive agricultural techniques.

Use of Pesticides The persistent use of pesticides, such as DDT, is another issue about which agricultural technologists and ecologists are at odds. Without question, pesticides have helped accelerate crop production and have improved the appearance of foods, yet we cannot evade the environmental consequences of their use. In 1962, Rachel Carson warned that the widespread and indiscriminate use of organochlorides would result in a decrease in the species diversity of our biosphere; this has happened. As the biohumanist Garrett Hardin emphasized: human beings cannot change only one thing. The "balance of nature" is a real concept, a corollary to the biologist's principle that there is no such thing as a unidirectional effect. The actions of human beings, aimed at increasing their own species' numbers, have often reduced or eliminated other species. As early as the Pleistocene Epoch, when human beings first appeared, small but efficient hunting bands may have caused the extinction of various mammals, such as mastodons.

From the standpoint of long-term gain, is the continued use of lethal pesticides justified? It has been shown that insect species quickly develop a tolerance to poisons. Due to their short generation time, resistant individuals surviving the poison quickly produce resistant populations. Then, different or increased quantities of poisons must be employed to control the surviving population. However, as the poison moves up the food chain, some vertebrates—especially birds and small mammals—simply die. Considering the prodigious amount of insects consumed by birds, agriculturists suffer an economic loss from their demise, if nothing else. The agriculturally beneficial effects of most pesticides, immediately impressive as those results are, may prove to be fleeting; whereas the destruction of a whole animal population is an irreparable loss.

Ethnobotany In many developing countries people live in the same way their ancestors have for generations past. They use plants just as their forebears did—not only for food, but for clothing, shelter, medicines, and pleasure. *Ethnobotany* is the study of the human uses of plants. As the world food crisis becomes of increasing concern to all, it is necessary to take advantage of a broader spectrum of potential foods. A variety of crops cultivated in ancient times and their uses have come under closer scrutiny. More attention is also being given to the medicinal plants of culturally primitive groups; many of these have been incorporated into the cultures because they possess some curative properties. There is renewed interest in ancient cures and remedies, potions, and foods. People in the developed countries are beginning to value the knowledge of cultural groups still using plants as an integral part of their lives in ways that have been taught to generation after generation for thousands of years. In the myths there are important grains of truth.

The Biosphere: A Delicate Balance All animal life is dependent upon plants. Thus plants are not only economically important to us; they are vital to our existence. They fix solar energy and replenish the oxygen supply necessary for animal life on earth. A deteriorating environment may result, at some future time, in more serious consequences than is popularly supposed. The pollution of the world's interconnected oceans, for example, may threaten the phytoplankton which produce about 70% of the oxygen in the biosphere. On land, crops grown near cities often suffer heavily; smog poisons the enzyme systems critical to their growth.

Food Chains and Energy Conservation Besides producing vital oxygen for the biosphere, plants supply the caloric base for all food chains.

Even a strictly carnivorous animal is entirely dependent for its existence upon the presence of plants. A food chain may be quite simple and direct—a cow eating grass; or it may be complex—in an oceanic chain phytoplankton are consumed by zooplankton, which are eaten by fish larvae, which in turn are eaten by small fish. Larger fish eat smaller fish, and so the chain progresses. A human being may eat the fish, ending the sequence.

Agriculture has brought about a number of artificial food chains, some of which represent calorically efficient usages of energy, and some of which are disturbingly wasteful. Conversion of energy is only about 10% on each level of a food chain. In general, human beings are at the end of the chain—gleaning their sustenance on numerous levels, from algae to mammals. In the future, we may have to effect a greater conservation of energy in providing solutions for feeding the world's human population. Planting crops yields far more energy, in the context of available land, than does raising cattle, sheep, or other domesticated animals. In order to feed our expanding human numbers, we must study how to achieve the optimum return from investments of land and resources.

Arriving at the most effective energy-conserving means for producing food is vital to the continuation of our highly-developed cultures, and to raising the standard of living in the developing countries. The level of efficiency from producer to consumer—including crop selection, crop improvement and innovation, crop use, storage and distribution, and efficiency of utilization—may be correlated directly with a society's standard of living and diversity and complexity of culture.

The decline of sophisticated civilizations in the past can be related to an expansion of the material culture and a population growth beyond the carrying capacity of the agricultural base. In many cases a companion element possibly was the sharp reduction of agricultural output due to overuse of the land. This happened to the Maya of Central America and the Khmers of Angkor Wat.

Plant resources currently in danger of drastic reduction or even depletion include extant ones, such as phytoplankton, and ancient ones in the form of oil, coal, and natural gas. The latter are all products of plants that lived millions of years ago. In predicting the future, some experts believe that the decline of world economics will be hastened by the depletion of both ancient and contemporary plant resources. The former, such as oil, exist in finite, exhaustible quantities. The contemporary resources—staple crops such as corn, wheat,

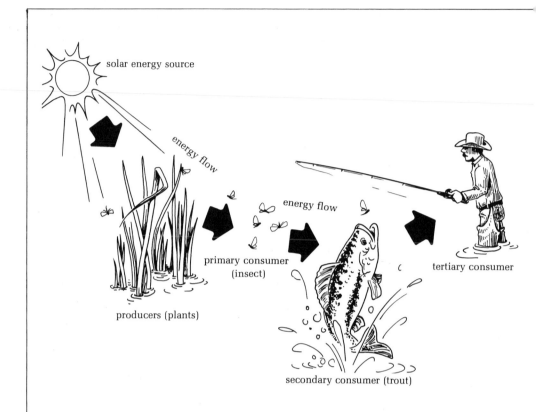

solar energy source

energy flow

energy flow

primary consumer
(insect)

producers (plants)

tertiary consumer

secondary consumer (trout)

Energy Flow

Solar energy source
Producers (plants)
Primary consumer (insect)
Secondary consumer (trout)
Tertiary consumer (human being)

and rice—may eventually lag too far behind the burgeoning world population. Also, in the interaction between human beings and their environment, staple crops are constantly threatened by human misuse of the land, such as may occur in the wake of extensive deforestation.

The optimistic perspective is that our high-yield, genetically-engineered crops will continue to stay famine from the thresholds of most of the world's nations. However, these crops have drawbacks with long-term implications. Typically, such crops require increased irrigation and more fertilizers. The manufacture of the fertilizers uses resources such as petroleum. More intensive cultivation from planting to harvesting is necessary; so more energy input goes into the cultivation. It is questionable whether these tactics in the war against famine will be affordable, in terms of the time and raw materials necessary. In the final analysis, the "success" of any biological interaction depends on a balance between the producer and consumer. To preserve a balanced and a continually productive ecosystem for not only the present generation but for those of the future, we cannot expect technological expertise to provide a remedy for the world food crisis when other dilemmas are ignored as they have been in the past. To continue feeding ourselves from the earth, we must redefine ourselves as a part of the biosphere, and acknowledge that we are subject to many of the same limiting factors as are other animals. Then we may make some progress.

This Course: A Combination of Studies As the world community grows smaller and the human population becomes larger, it is no longer functional to separate the studies of botany, anthropology, demography, and economics. We cannot effectively consider problems in agriculture without weighing other related components of the entire question. Political, economic, and transportation factors may have as much or more to do with feeding a population than the actual quantity of food available. Food distribution and the cost to the individual must be taken into account.

The traditional scope of courses focusing on plant and human interaction has been little more than the study of commercial crops and the control of insect pests. This was a valid approach in a less complicated era. Today, this course must deal with a range of complex and crucial problems. The first of these—at least from a humanistic standpoint—is how to assure at least an adequate diet for the world's population. The solution rests not only with more efficient agriculture, but with population control measures, as well. Yet even with the most successful birth control programs, there will be many millions more to feed. Unless it is possible to break down nationalistic barriers, to cooperate on a worldwide basis to produce more food, numerous millions are destined to starve.

A merger of economic and humanistic disciplines with the study of human interactions with plants is sensible in other contexts as well. Human beings have been "cultivating" plants in peripheral ways long before the establishment of a stable agricultural tradition. Primitive peoples dispersed the seeds of edible varieties, planted a few individuals of desirable types near their dwellings, or perhaps cleared out extraneous kinds to allow useful plants more growing space. There is no legitimate reason for relegating these historical aspects of the study to one discipline—anthropology—while treating contemporary agriculture as a topic of separate importance.

Over the centuries, human cultural evolution has proceeded integrally with the use of plants; it has been a continuing process and should be treated as such. In this book we have tried to do so.

CHAPTER

2

CLASSIFICATION, STRUCTURE, AND REPRODUCTION OF THE ANGIOSPERMS

Classification   The classification, or arrangement into groups, of plants and animals is both an ancient and a current activity. In modern biology, *taxonomy* is the system of arranging plants and animals into related groups. Such groups are based on common factors in structure, embryology, or biochemistry.

However, naming and classification of plants was done long before modern taxonomy. Preagricultural peoples surely possessed an extensive knowledge of plants in terms of a variety of qualities. They knew whether or not the common plants were edible, toxic, curative, or "magical," as evidenced today among extant societies of non-agricultural or minimally agricultural peoples. Often their understanding rivals that of our modern systems of classification. Their perspective and accumulation of plant lore is a necessary aspect of their way of life. Several years ago one of the authors was collecting ethnobotanical information from Carib Indians living in the region of Dawa in Guyana, South America. From a list of 50 plant species described by Western science, the Indian guide supplied 50 distinct native names. Among such people, a knowledge of local flora is a part of survival. Most of the people in the village, including children, were able to identify correctly 95% of the local flora by their native names. In contrast, the average citizen of an industrialized society might be pressed to identify all of the produce in a supermarket!

From Aristotle to the Present   The study of plant classification as a discipline of Western science began with the work of the famous Greek philosopher and naturalist Aristotle and his pupil Theophrastus (370–287 BC). Together they compiled the oldest Western botanical work in existence, the *History of Plants*. Theophrastus described some 500 species of plants, most of them cultivated varieties, and classified them as herbs, shrubs, undershrubs, or trees. His work involved fundamental botanical classification, which is the assembling of plants into groups (*taxa*) on the basis of their assumed relationships.

Classification can be done using any arbitrary criteria, such as edible vs. nonedible; woody vs. herbaceous; or economic usefulness vs. economic irrelevance.

A Natural Taxonomy   Contemporary botanists attempt to classify plants by "natural" rather than by arbitrary systems. A natural taxonomy attempts to define relationships on the basis of genetic and evolutionary characters. Some highly similar species are, in fact, not related but have developed like structures in response to equivalent environmental demands. This means of classification requires a far greater knowledge of plants and entails much more detailed study than the simple process of designating a common name—such as blue-grass, turkey-oak, or sunflower. In modern botanical taxonomy, relationships must be determined through the study of many nonevident characters. The academic taxonomist would be sorely hindered without the aid of technological apparatus such as the microscope.

In assessing and establishing degrees of relationship among plant groups, categories accepted and established by the International Code of Botanical Nomenclature provide a basis for more or less uniform agreement (see page 24). Seldom are all of the available categories filled in. Factors to be considered include the size of the group of plants, how extensively they have been studied, and the complexity of evolutionary mechanisms operating within the group.

Carl von Linné (Carolus Linnaeus, 1707–1778), is generally considered to be the father of modern plant classification. He first recognized some of the currently accepted categories: class, order, genus, species, and variety.

Defining a Species. The species is the basic unit of formal classification. It is the only one of the many taxonomic categories that has a biological reality in nature. That is, a biologist in the field can collect *individuals* of a species, and also of the species' subgroupings: subspecies and variety. However, a botanist cannot collect a family, or a genus. These are abstract categories used to order the "real" categories of species, subspecies, and variety. Species are recognizable in nature because the biological variation that defines them is more or less continuous within the group. To put it another way, species' characteristics are discontinuous from one species to another.

Biologists define a species by using two basic concepts: one morphological and the other biological. Morphology refers to structure, and morphological definition focuses on external characteristics. Basically, morphological classification depends upon an analysis of identifiable characters, such as form and coloration. This is a valid means for imposing order upon unknown groups of organisms. However, a taxonomy based upon external features alone might bypass important relationships not visually obvious, or relationships might be assumed when there were none.

Certainly traditional morphological taxonomy has its purpose, but its accuracy is limited; whereas the total biological description of a species attempts to define a complete relationship derived from morphological, cytological (biology of cells), and genetic characteristics. Even with the biological

International Code of Botanical Nomenclature (1961). All ranks need not be used, but the order may not be changed.

Divisio (*Spermatophyta*)
  Classis (*Angiospermae*)
    Subclassis (*Dicotyledones*)
      Ordo (*Rosales*)
        Subordo (*Rosineae*)
          Familia (*Rosaceae*)
            Subfamilia (*Rosoideae*)
              Tribus (*Potentilleae*)
                Subtribus (*Potentillinae*)
                  Genus (*Potentilla*)
                    Subgenus
                      Sectio (*Drymocallis*)
                        Subsectio (*Closterostylae*)
                          Series
                            Subseries
                              Species (*Potentilla glandulosa*)
                                Subspecies (*P. glandulosa* ssp. *nevadensis*)
                            Varietas
                              Subvarietas
                          Forma
                          Subforma

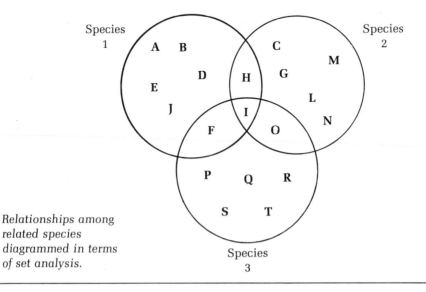

Relationships among related species diagrammed in terms of set analysis.

definition, the demarcations between species may, at times, prove to be illusory. However, an approach taking many factors into account is more accurate than the old-fashioned morphological approach alone.

In terms of set analysis (see page 24), Species 1 contains elements A, B, D, E, and J, which distinguish it from Species 2 and 3, both of which have their own distinct elements. Note that elements H, O, and F are common to two of the three species, and element I is shared by all three.

It is often true that morphological taxonomy will concur with the findings of more analytical biological examinations. Both avenues of investigation should be allowed a valid place in modern botany.

Subspecies. Subspecies have a biological reality usually more difficult to detect than that of a species. The subspecies concept was used widely during the latter part of the 19th century. The concept was often misapplied because the causes of variation within species were not adequately understood, and the range of variation within a single species was not recognized. Today, a *subspecies* is defined as a population, within a given species, which is not isolated in its gene flow from other populations of the same plant (or animal); yet which exhibits a series of predictable characters distinct from other individuals in the species (taxon). Given conducive circumstances, such as genetic isolation from the parent group, a subspecies may evolve into a full species.

Variety. The term *variety* refers to a local variation in a species' natural population, or to artificially produced varieties resulting from the directed hybridization of plants. The term indicates some degree of morphological distinction that sets the group apart. However, the distinction is not broad enough or various enough to be called a subspecies. Agricultural varieties that are the products of artificial selection by human beings are known as *cultivars*. A variety or cultivar has only a few distinctive characters, usually applicable to the growth form, flower, or fruit in cultivated varieties.

Many taxonomists feel that the subspecies is the only intraspecific (within the species) category worth recognizing. But this may depend upon the size of the group and the extent of its economic utilization.

Genus. Interspecific categories are those established between and among species. Essentially, these are constructions created for the understanding and convenience of the taxonomist. The *genus* is the basic division for grouping species and functions to delineate species with a genetic affinity. Number of species is often a consideration in establishing a genus. For example, if there are only two or three species exhibiting characters intermediate between two genera, it may be preferable to leave those species in one of the two existing genera until the whole complex is better understood. If new factors are discovered, the taxonomist might then create a third genus to accommodate those several species. On the other hand, if there were 75 species with characteristics intermediate between those same two genera, a taxonomist might create a third genus more readily.

To the disgruntlement of some taxonomists, many closely linked genera show little, if any, clearly distinguishable differences between them. This happens more often with plants than with animals. Generic characters, most often defined by floral characteristics, tend to represent environmental adaptations. Such characters are the outcome of the stresses of natural selection. Occasionally, however, the phenomenon of *genetic drift* might occur. In these cases, neutral qualities become fixed in a population, regardless of the adaptive value of the character. Genetic drift may be responsible for certain predictable features in a population.

Family.   The goal of the taxonomist is to define families that are natural by reflecting a relationship at the genetic level. Flower and fruit structures yield the most useful indications of family groupings. However, it is not always possible to define large groups accurately. Definable families are clearly delineated and reflect genetic affinities among their generic components.

Order.   The taxonomist is most challenged on the level of ordinal classification. An *order* of plants is a cluster of closely related plant families. The criteria traditionally used to define an order is (1) the position of the ovary in relation to the floral parts, or (2) the internal partitioning of the fruit and the placement within the fruit of its seeds. With flowering plants (angiosperms), the order is the least understood of the taxa employed in classification. The result of this confusion with angiosperms has been an arbitrariness in making up orders, and the creation of a great many of them.

The Naming of Plants   How do all of these categories apply to the plants themselves? Each plant has two names: a specific name and a generic name. This two-part scientific name links the individual plant's species with others to which it seems to be closely related. For example, in the case of common wheat the specific label is *aestivum* and the generic is *Triticum*. In combination the names are written *Triticum aestivum* L. The "L" indicates that the species was first described by Linnaeus. (Among more recent investigators, the botanist's whole name is customarily given or abbreviated). Let's examine a hypothetical situation in which the wheat taxon to be studied is given a new name, *arvense*, by a hypothetical person named Long. The new appellation would be written *Triticum arvense* (L.) Long. In this book, when a genus has been discussed previously, the generic name will be subsequently abbreviated to an initial: *T. aestivum* L. Usually, the designation of the investigator is dropped, so the species would be referred to merely as *T. aestivum*.

A complete categorization of common wheat, *Triticum aestivum*, is as follows:

Kingdom—Planta
Division—Spermatophyta
Class—Monocotyledonae
Order—Graminales
Family—Gramineae
Genus—*Triticum*
Species—*aestivum*

Structural Elements of Angiosperms   In order to understand the terminology that will appear in later chapters, it is necessary to consider the structural and reproductive features of the angiosperms. This is the group of plants of most concern to man; most of our major food crops are members of this class.

The Cell Wall   One of the most critical features distinguishing plant from animal cells is the rigid cell wall, which is also the source of many economically important plant products. The *cytoplasm* is a colloidal substance constituting the living matter of the cell. The cytoplasm secretes the plant's cell walls. There is a primary wall, and usually a secondary wall. The walls may have three components: *cellulose, lignin,* and *pectin,* which are chemical polymers involving different linkages of molecules and auxiliary elements. A *polymer* is a compound consisting of repeated linked units, each a relatively simple molecule. *Cellulose* is a polymer of glucose, a simple sugar. *Hemicellulose,* a polymer found in some plants in addition to cellulose, contains both sugars and non-sugars. *Lignin* contributes to the woodiness of cell walls. It consists of polymers of phenyl propanoid units. *Pectin,* the third cell wall component, is formed from water-soluble polymers of galac-

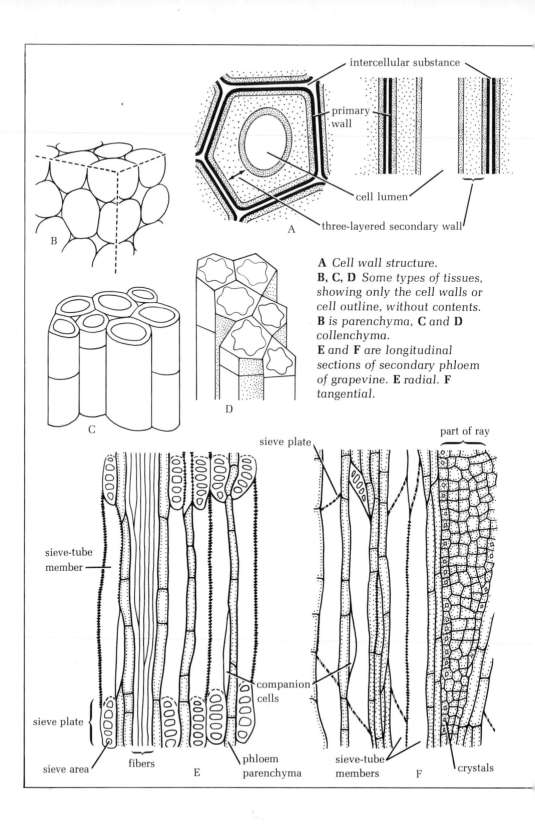

intercellular substance

primary wall

cell lumen

three-layered secondary wall

B

A

A Cell wall structure.
B, C, D Some types of tissues,
showing only the cell walls or
cell outline, without contents.
B is parenchyma, C and D
collenchyma.
E and F are longitudinal
sections of secondary phloem
of grapevine. E radial. F
tangential.

C

D

sieve plate

part of ray

sieve-tube
member

companion
cells

sieve plate

sieve area

fibers

E

phloem
parenchyma

sieve-tube
members

F

crystals

turonic acid. It is important because of its ability to form sols and gels in water.

The cell wall itself may be secreted in two distinct layers. The primary cell wall is formed first, of cellulose and pectin. The secondary cell wall is formed on the inside of the primary wall. In addition, pectic substances form a thin layer between the primary wall of two adjacent cells. This layer, called the *middle lamella*, glues the plant cells together. Depending on the cell type, lignin may or may not be formed. If lignin is produced, it may infiltrate the middle lamella, the primary wall, and the secondary wall.

Tissue Types
Cells are the basis of *tissues*—continuous organized masses of cell types. In a mature plant, there may be three types of simple tissue systems; each system is composed of tissues with only a single kind of cell. These systems are *parenchyma*, a tissue formed from simple, thin-walled cells; *collenchyma*, a supporting tissue comprised of thick-walled cells; and *sclerenchyma*, a thick-walled tissue, often "stony" in texture, as in the shell of a walnut. In addition to these three simple tissues, there are two kinds of complex tissues consisting of more than one cell type. Two important complex tissues are: *xylem*, the water-conducting tissue which is the wood of woody plants, and *phloem*, the food-transporting tissue (see page 28). These tissue types are the basis of various economically important plant products. *Parenchyma* forms the fleshy part of fruits and roots. If chlorophyll is present in the parenchyma, then photosynthesis takes place in this tissue. *Photosynthesis* is the use of light energy by the plant to synthesize carbohydrates from carbon dioxide and water. *Collenchyma* tissue is a mass of elongated cells which function primarily as a mechanical support for the growth of young stems and leaves. An example of this tissue is the "threads" of a celery stalk. *Sclerenchyma* tissues are throughout the plant, frequently nonliving when mature, and mostly lignified. These tissues may form as sheets of cells, giving rise to economically important fibers; or they may be diffuse, an example being the grit of a pear fruit. Often, sclerenchyma is quite dense, as in the shells of nuts. Whereas sclerenchyma tissue is important to fiber production, not all fibers are formed from it. Wood itself, consisting of xylem, may be classified as fibrous. Bast fibers are associated with phloem. One of the most important of all commercial fibers—cotton—originates as hairs growing from epidermal cells of the seed coat.

The Root and the Shoot From its beginning as an embryo in the seed, two morphological regions may be distinguished in a vascular

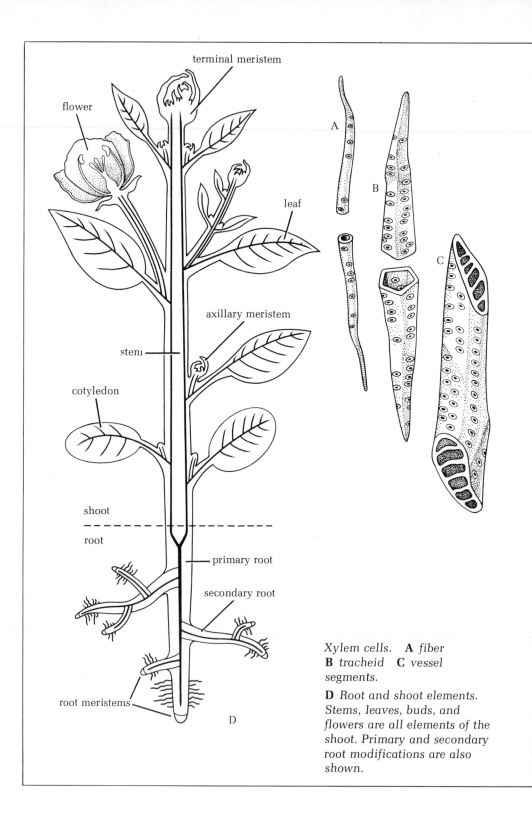

terminal meristem

flower

leaf

axillary meristem

stem

cotyledon

shoot

root

primary root

secondary root

root meristems

D

A

B

C

Xylem cells.   **A** *fiber*
B *tracheid*   **C** *vessel*
segments.

D *Root and shoot elements.
Stems, leaves, buds, and
flowers are all elements of the
shoot. Primary and secondary
root modifications are also
shown.*

plant: the root and the shoot. Stems, leaves, buds, and flowers are all elements of the shoot. These elements appear in a variety of structures conveniently classified as above- and below-ground modifications.

Shoot Modifications. Above-ground modifications include crowns, which are compressed stems; cabbages are representative. Stolons are horizontal stems highly adapted to the production of shoots and roots from buds that lie along their length. Their capacity for vegetative reproduction is typified in strawberry plants. Spurs, short stems specialized for flower and fruit production, are found in many flowering trees, such as the apple.

Below-ground shoot modifications are storage regions, and are economically important due to the energy concentrated in them. The "bulb" of an onion or a tulip is really an underground stem wrapped in a number of leaves which thicken considerably at the base. The layers of an onion are merely the bases of these leaves. A corm, exemplified in a crocus, is an extremely short and compressed underground stem. A rhizome, found in an iris, is a horizontal underground stem. A tuber, such as a potato, is an enlarged portion of a short, fleshy, underground stem. Many structures that appear to be part of the root system are, in actuality, portions of modified shoots.

Leaves and Flowers. The most conspicuous manifestations of a plant's shoot are usually its leaves and flowers. Leaves constitute the principal photosynthetic organ. Flowers, although distinctive in appearance from the leaves, have their origin as modifications of the leaf structure. They are a reproductive segment of the shoot system, with specially modified appendages. The end of the branch supporting a flower is the receptacle, and the leafless area directly below it is the pedicel. (See the illustration on the facing page of a generalized flower with major parts labelled.)

Reproduction in the Angiosperms Plant reproduction is the real basis for agriculture in the sense of human beings' taking advantage of all plant reproductive processes—both sexual and asexual. In the most general terms, reproduction is the total sequence of events required for the replication of cells and organisms.

Sexual Reproduction In the biology of the angiosperms, sexual reproduction involves the formation of male and female gametes through the process called reduction or meiosis. Reduction involves the division of the plant cell's chromosomal complement to form gametes. Typically, a gamete has half the number of

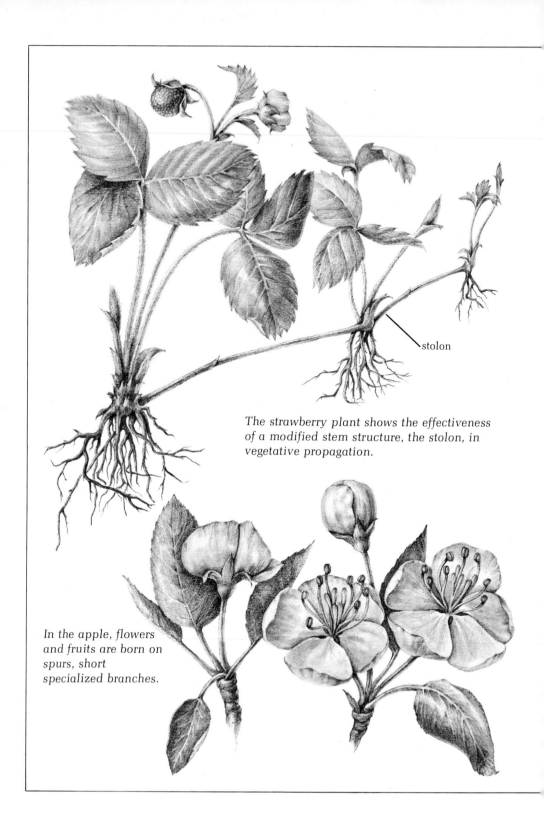

stolon

The strawberry plant shows the effectiveness of a modified stem structure, the stolon, in vegetative propagation.

In the apple, flowers and fruits are born on spurs, short specialized branches.

chromosomes needed to produce a new individual. Reproduction is the subsequent fusion of these gametes, each containing one-half of the genetic material of the parent plant, to reconstitute the full complement in the fusion product: the zygote. Through subsequent cell division and differentiation, the zygote then grows into a mature plant.

Nearly all of the angiosperms are *diploid* in their chromosomal complement, that is, each cell of the plant contains *two* possible alternate forms of expression for each particular hereditary character. Gametes contain only *one* possible expression of a hereditary character. Thus a gamete has half of the genetic complement of a normal cell and is said to be *haploid*. These alternative forms of expression for a single gene (or character) are called *alleles*. Changes in the chromosome combinations of alleles making up the total genetic complement of a plant can cause major variation in physiology, morphology, and structural adaptation. We observe these plant changes about us all the time; they are called *mutations*.

A peculiarity of plant reproduction usually not observed in animals is that plants often duplicate whole or incomplete sets of their genetic complement. For instance, a plant cell may have *three times* the number of chromosomes found in the gamete. This phenomenon, *polyploidy*, is important not only to the genesis of new varieties of cultivated species of plants, but to the occurrence of new species in nature as well. Let us look at a specific example of *polyploidy*. A diploid plant might have a chromosomal complement of 10 complete sets of genetic information, or chromosomes. Through failure of mechanical separation of chromosomes during the meiotic process, due either to natural or artificial agents in the plant's environment, the gametes formed may contain the entire genetic complement of the parent plant. In this case that would be 10 chromosomes rather than the half (in this case 5) typically appearing in the gamete.

When a gamete with the entire parental complement fuses with a gamete that has the usual half complement—three sets of chromosomes arise. It is then possible to derive a functional fertile progeny having a total chromosomal complement of 15. If two gametes were formed without the usual reduction of either (in this case 10 + 10), the progeny would contain 20 complete sets of chromosomal information. Such instances are examples of *autopolyploidy*, the origin of whole self-replicated sets of chromosomal information within the parent plant.

It is not only possible, but occurs frequently that a complete chromosomal set from another closely related species may be

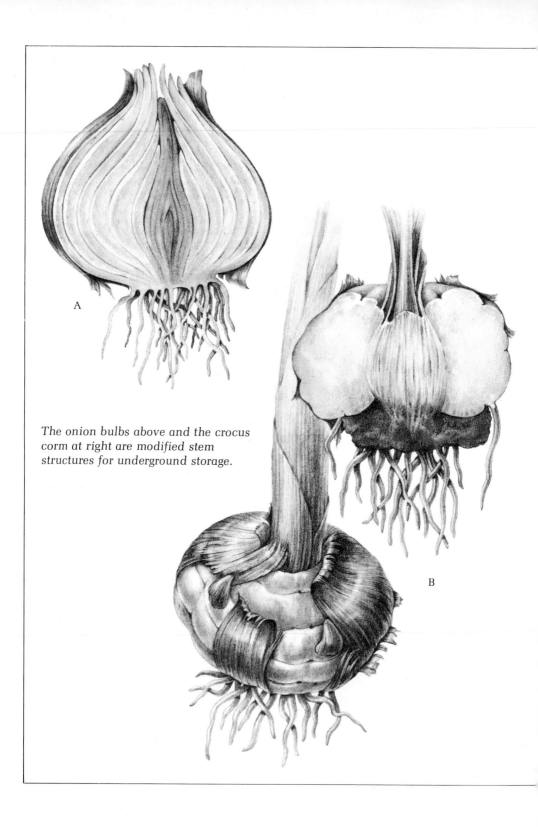

The onion bulbs above and the crocus corm at right are modified stem structures for underground storage.

A

B

The iris rhizome at left and the potato above, a tuber, are also underground stem modifications.

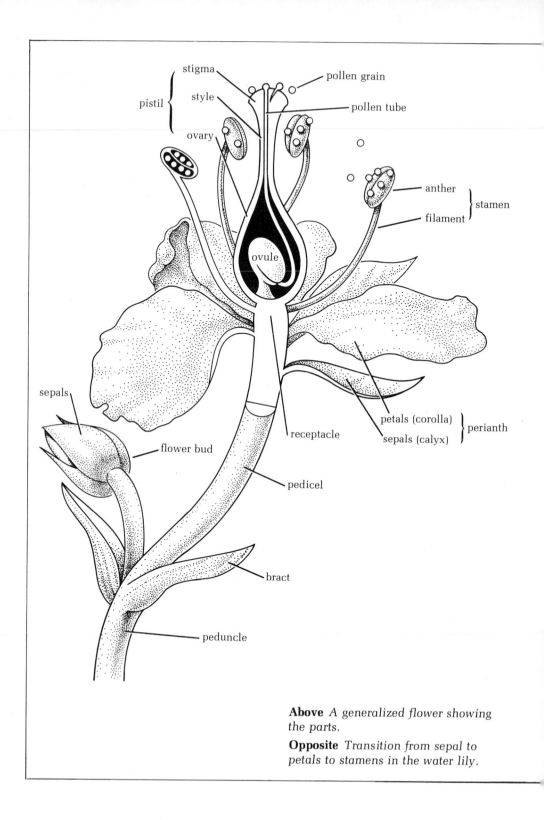

Above *A generalized flower showing the parts.*

Opposite *Transition from sepal to petals to stamens in the water lily.*

added to the haploid set (*genome*) of a plant by hybridization. For example, if a given plant with a diploid complement of 10 forms normally reduced haploid gametes (5 chromosomes per gamete), and if one of these gametes were to fuse with a nonreduced gamete of a closely related kind in which the usual chromosomal complement is 12—it would be possible to produce a progeny with a complement of 17. This condition is referred to as *allopolyploidy*.

Sometimes, whole chromosomes may be lost during meiosis, thus being excluded from the formed gamete. In our example of the diploid species having 10 chromosomes, a gamete with 3 instead of 5 chromosomes might be formed. Its fusion with a normal gamete having a complement of 5 sets would then result in offspring having a total complement of 8, a condition known as *aneuploidy*.

Asexual Reproduction It should be noted that vegetative, or asexual reproduction is exceedingly common—not only among those species of plants which have lost the capacity to effect sexual reproduction, but among those using sexual reproduction as a principal strategy. Asexual reproduction may be thought of as an additional degree of freedom open to a reproducing plant, enlarging the plant's whole reproductive potential. In circumstances making sexual reproduction impossible, asexual reproduction becomes a temporary means of survival.

The most significant advantage of sexual reproduction over asexual is that the former process allows for more potential variation within a species. This is because in sexual reproduction all possible allele combinations can be subjected to "environmental testing" by a species in its pursuit of biological survival. A new mutation may predominate; old genetic material may be deleted. However, sexual reproduction

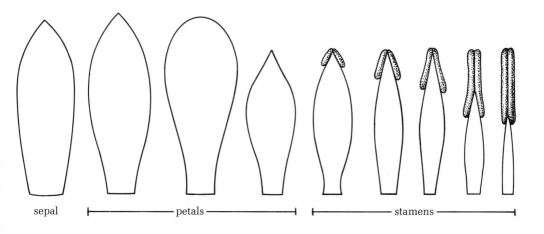

sepal |—————— petals ——————| |—————— stamens ——————|

Life history of a flowering plant.

A *Ventral view of the anther.* **B** *Cross sections of the anther before (above) and after (below) dehiscence.* **C** *Flowering plant pollen grain development.* **1** *Microspore developing into a pollen grain.* **2** *Pollen grain composed of two cells, one within the other.* **3** *Pollen grain germinating in a sugar solution, the pollen tube emerging.* **4** *Pollen tube with the male gametes (nuclei).*

D *Flower stage of the pistil with the side cut away, showing the ovules.* **E** *Young ovule in side view.* **F** *Ventral view of a young ovule.* **G** *Longitudinal section, with pollen grain approaching from above.* **H** *Fertilization: one male gamete nucleus joins the egg nucleus, the other the proendosperm nucleus.* **I** *After fertilization.* **J** *Development of the embryo from the fertilized egg, and development of the endosperm.*

K *Fruit (a pod or legume).* **L** *Fruit with half of the ovary wall removed.* **M** *Seed, external view, and sectional view showing the embryo.*

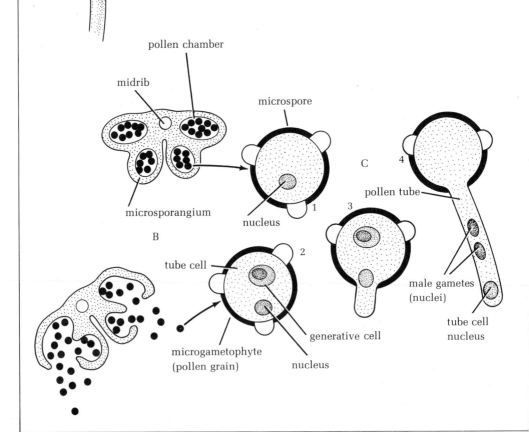

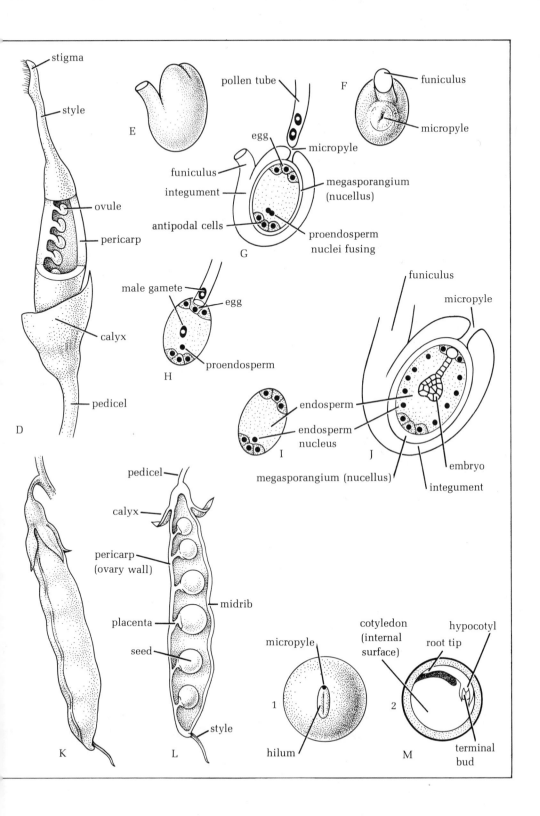

stigma

style

E

pollen tube

F — funiculus

micropyle

ovule

pericarp

egg
funiculus
integument
antipodal cells

micropyle
megasporangium
(nucellus)
proendosperm
nuclei fusing

G

calyx

male gamete
egg
proendosperm

H

funiculus

micropyle

pedicel

endosperm
endosperm
nucleus

embryo

D

I

J

megasporangium (nucellus)

integument

pedicel

calyx

pericarp
(ovary wall)

midrib

placenta

seed

style

micropyle

hilum

cotyledon
(internal
surface)

hypocotyl
root tip

terminal
bud

K

L

1

2

M

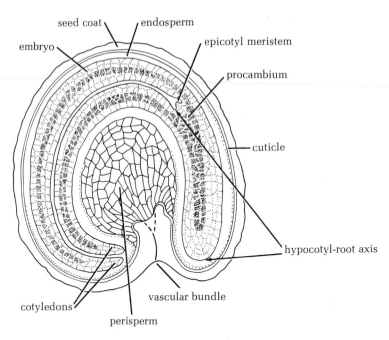

seed coat · endosperm

embryo

epicotyl meristem

procambium

cuticle

hypocotyl-root axis

vascular bundle

cotyledons

perisperm

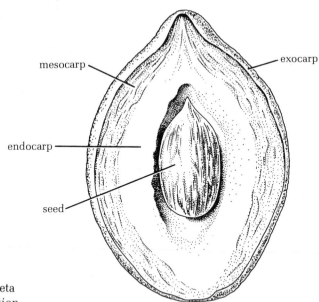

mesocarp

exocarp

endocarp

seed

Above *Seed of sugar beet (Beta vulgaris) in longitudinal section.*

Right *A fruit is a ripened ovary. Longitudinal section of a young ovary of almond (Prunus amygdalus) showing the pericarp surrounding the seed.*

has been lost in many plant species, or it may be non-functional due to the absence of a pollinator. This happens to cultivated varieties far removed from their natural habitat, or to plants growing under stressful natural conditions.

Plant Propagation The Use of Seeds. The use of seeds is probably the oldest and most dependable method of *plant propagation*. Replication of plants is possible by both sexual and asexual means. During the origins of agriculture, particularly in areas with climates which restricted the growing seasons, a reserve of seeds held from one season to the next was critical to crop production, and to human survival. Seeds are still one of the most common and reliable means of plant propagation for several reasons. First, they are inexpensive, relative to the cost of germinated seedlings or rooted cuttings. Second, they are easily stored, and a high percentage of the seed lot will remain viable for germination from one year to the next. Lastly, seeds may be treated with fungicides and other microbiocides, although it is possible for some virus and genetic diseases to be carried inside the seed.

Some plants, however, do not breed true from their seeds because of high rates of genetic recombination. With plants which do not duplicate from seeds, asexual or vegetative propagation must be employed. In the case of potatoes, for example, "seed potatoes" are planted. These are cut sections of a potato, including one or more buds.

Vegetative Propagation. Vegetative propagation may be an artificial process, as in the above example, or it may be an integral part of the reproductive tactic of a sexually reproducing plant. Strawberry plants, which set viable seed, also propagate themselves through the production of horizontal stems, called runners. These runners contain buds that develop into roots and shoots under favorable growing conditions. Vegetative propagation has several advantages in the cultivation of crop plants. First, those plants which do not breed true to the parent plant can be reproduced without genetic changes such as recombination. Second, vegetative propagation is frequently an easier and sometimes a faster means of cultivation than the use of seeds to establish a crop. Lastly, it is essential for the perpetuation of seedless varieties of bananas, pineapples, oranges, and other fruits.

Several techniques may be employed in vegetative propagation to assure that the desired qualities of the parent plant will be passed on without change. One of the most curious of

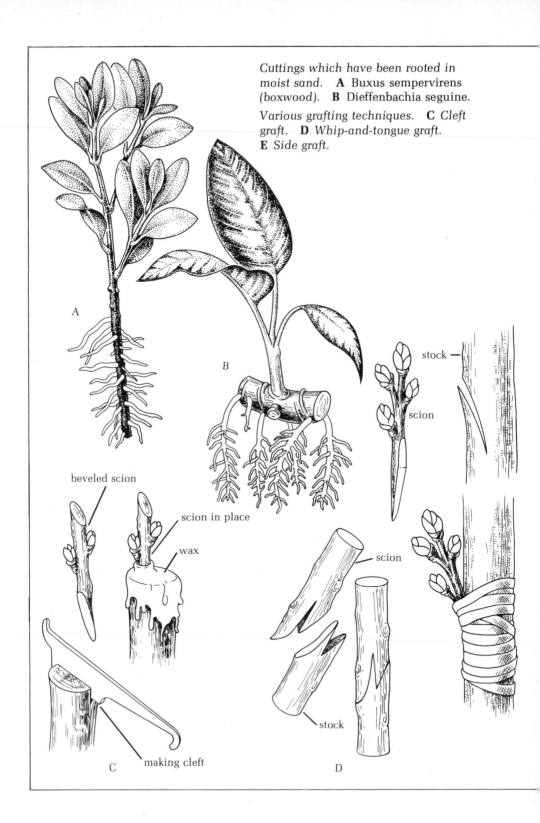

Cuttings which have been rooted in moist sand. **A** Buxus sempervirens (boxwood). **B** Dieffenbachia seguine.

Various grafting techniques. **C** *Cleft graft.* **D** *Whip-and-tongue graft.* **E** *Side graft.*

A

B

stock

scion

beveled scion

scion in place

wax

scion

stock

making cleft

C

D

these is the use of seeds that have arisen *without fertilization* and contain only the maternal chromosome sex. Such *apomictic* seeds have been used occasionally in the cultivation of some citrus fruits and mangos, and regularly for Kentucky Blue-grass.

Tissue Culture.   A laboratory technique developed recently for the asexual propagation of plants is tissue culture. Cells from the parent plant are grown in a nutritive medium containing factors necessary for the cells' development into whole plants. The technique has been adopted by many commercial operations, such as in the growing of orchid plants. In the case of orchids, the time interval between seed germination and flowering normally encompasses several years. Using the tissue culture method, flowering plants can be obtained frequently in less than half the normal time period required for development; and with tissue culture the retention of desired characteristics is assured.

By far the most common method in commercial vegetative reproduction is that of using structures produced by the plant itself. These are most often specialized stem structures: runners, bulbs, corms, sections of the above-ground stem, and underground stems, such as tubers and rhizomes. The induction of roots in a place where roots would not usually appear (such as the plant stem) is possible by the use of cuttings or by air layering (see A and B). In the latter process, the bark of woody shrubs or trees is cut away from a section of stem around which sphagnum moss is packed. Root-forming hormones may be added to stimulate growth, and the whole section is then sealed with water-impermeable material, usually a plastic tape or wrap. Within several weeks, roots begin to form; and the shoot portion can be removed from the parent plant, to be potted with its own root system.

Grafting.   Through the process of *grafting*, portions of the shoot system of one plant can be removed and incorporated into the shoot system of another plant of the same species or of a closely related species (see C thru E). The total product of this technique is a *graft*. The plant part which provides the root system is the *stock*. The grafted shoot is the *scion*. Many varieties of plants are grown with the aid of grafting, which offers diversity in the shoot system when space for the plant itself is limited. In a small garden, several kinds of apples may be grafted on the same root, or several kinds of roses thrive on a single bush. Grafting has commercial advantages. Certain citrus varieties exhibit different properties in the shoot or root systems in relation to productivity, disease resistance, or other qualities. It is, therefore, often profitable to graft the branches of one variety which bears prolifically onto another variety with a hardier root structure more resistant to insect pests and plant diseases.

Most such asexual plant propagation requires the formation of roots from some part of the shoot. As a process, rooting involves the division of cells and the cells' subsequent differentiation. Thereafter, plant growth is the sum of cell division, cell enlargement, and cell differentiation. Plant cuttings and layerings in both woody and nonwoody plants form roots initially in the vicinity of vascular cambium, the growing part of the plant that produces the xylem and phloem. The process of rooting is physiologically complex, with specific problems and considerations in each situation. Yet particular variables clearly affect the process: the amount of reserve food in the plant determines root formation, on a more or less direct level (large amounts increase it); exposure to light usually inhibits root growth; specific hormones, such as IAA (indole acetic acid) promote it. Some woody plants will not root when they are mature, and must be "juvenilized" by repeated pruning of the shoots. The cuttings will then root.

Suggested Readings

Benson, L. *Plant Classification*. Boston: D. C. Heath & Company, 1957.

Cronquist, A. *Introductory Botany*. New York: Harper, 1961.

Esau, K. *Plant Anatomy*, second edition. New York: John Wiley & Sons, 1964.

Sinnott, E. W. *Plant Morphogenesis*. New York: McGraw Hill, 1960.

CHAPTER

3

EVOLUTION, COEVOLUTION, AND THE ANGIOSPERMS

The angiosperms are now the dominant component of the world's flora. How did this class of plants triumph over other organisms?

Evolution: The Result of Biofeedback Evolution may be thought of as the result of the total biofeedback from the biological and physical environment on an organism over time. This feedback affects the organism's genetics, structure, physiology, and interactions with other organisms. The biofeedback effect is continuous; thus a species cannot be static. The biological and physical world is dynamic. Organisms must interact continuously with the changing environments or become extinct. It can be said that all organisms are involved in dynamic change with their environment.

In considering evolution, there may be a tendency to think in terms of some guiding force. This is a misconception—distorting the reality of the descriptive nature and mechanics of the evolutionary process. Presumably no organism or organismal interaction evolved toward its present complexity by design. What an organism is and what that organism may be lies within a parameter of the genes of the organism and the continuous biofeedback from the environment. The origin and the success of a feature of an organism's design is the result of chance interaction between genetic and environmental factors. Let us look at an analogy. Consider the process of evolution as a gambler who attempts to survive by betting (natural selection) on the best species. The chance of a pay-off is determined by the effectiveness of an organism's biofeedback interaction between its genetics and its bio-physical environment. Regardless of the actions of the gambler, the outcome of the species is determined by chance.

Natural Selection The most popularized aspect of evolution as conceived by Darwin is the process of *natural selection*. The term refers to differential reproduction among individuals of a species—resulting in varying responses of the species' population segments to the sum of environmental pressures. Such a response can be measured as a change in the frequency of inheritable traits in the population segments. From an evolutionary perspective, the adapted individual or population segment of a species is the one leaving the most progeny to contribute to the future genetic composition of the species and the survival of the species. Within the scheme of natural selection, the segments of a species that accommodate to changing conditions most effectively leave the most offspring and thus have the larger part determining the genetic make-up of that species in the future. But natural selection cannot induce evolution. There must be mechanisms for changing the genetic information of a species passed from one generation to another. These mechanisms are mutation, recombination, and isolation.

Mutation

New genetic information arises in individuals of a population. One of the most common mechanisms producing changes in genetic programming is *mutation*. Mutation is a change in the genetic information imparted to cells which acts to alter the transmission of hereditary instructions. Mutations occur constantly in populations of species. The occurrence may be spontaneous or the result of many environmental inducing agents. The great majority of mutations (99+%) are harmful or productive of irrelevant changes at any given point in evolutionary time. However, the small percentage that are not harmful or irrelevant may endow a population with beneficial characteristics. Mutations are recurrent —allowing a population segment of a species a continuous supply of genetic degrees of freedom to solve the problems presented by a changing environment, through a potential increase in genetic variability.

Recombination

Individuals contribute to the genetic variation of their species in yet another way: through *recombination*. By mixing particulate units of heredity (genes) or unit sets of hereditary information (chromosomes), a variety of new combinations of hereditary information is produced. These changes are the result of the remixing of existing units, whereas mutation allows the addition or deletion of "new" information.

Hybridization. *Hybridization,* or the crossing of genetic complements from different populations, is another facet of recombination. It is of the utmost importance to human beings in their interaction with culturally and economically important plants. A species undergoing hybridization with another population segment of that same species—or often with a segment of another species—has a genetic complement that allows it to survive under the selective pressures of its unique environment. The species' very survival indicates that those genes are adaptive to that particular environment. Hybridization involves the mixing of two or more adaptive gene complements. Often, the result is an increased potential adaptation of the new genetic entity, the *hybrid.* The survival value of the hybrid is increased and, often, its usefulness to human beings.

Hybrids—like mutations—are frequently less adaptive than the parent plants. A few hybrid progeny, perhaps 1 in

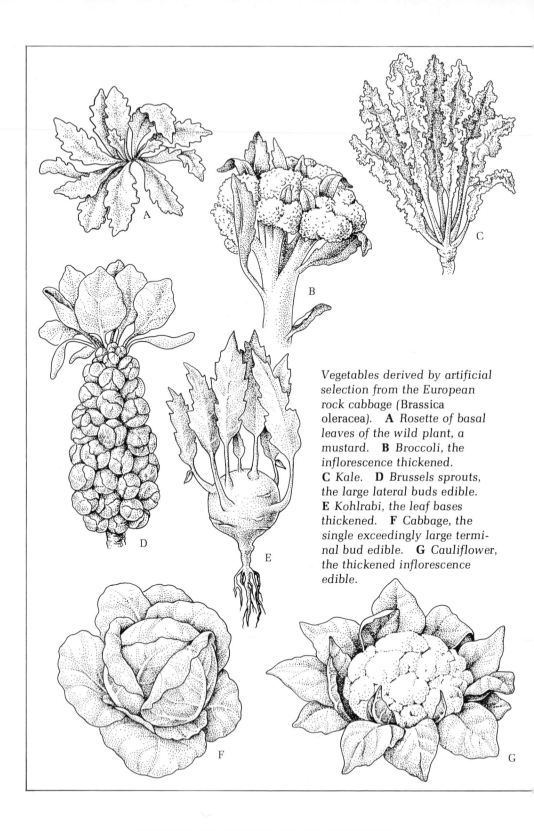

Vegetables derived by artificial selection from the European rock cabbage (Brassica oleracea). **A** Rosette of basal leaves of the wild plant, a mustard. **B** Broccoli, the inflorescence thickened. **C** Kale. **D** Brussels sprouts, the large lateral buds edible. **E** Kohlrabi, the leaf bases thickened. **F** Cabbage, the single exceedingly large terminal bud edible. **G** Cauliflower, the thickened inflorescence edible.

100 or 1 in 1000, possess an overall superiority to both parents. Under natural selective conditions, these hybrid progeny usually show a reproductive advantage, contributing more to the genetic complement of future generations. In the context of artificial selection by human beings, hybrid plants form the base for most important and useful crops. Today, there are few crops or animal stocks that are not the result of some conscious hybridization, whether the process took thousands of years or only a few aided by our recent knowledge of genetics. Early agricultural humans unknowingly hybridized both plants and animals by choosing desired individuals for establishing new stocks and crops. Contemporary scientists do the same, aided by a vastly greater reservoir of information.

Adaptation as Response to Natural Selection   Populations of a species respond to the pressures of natural selection by specific adaptations which can be measured in the frequency of heritable characters in that particular population. Such adaptations might be enzyme differences, physiological features, or morphological discontinuities. Such genetically based differences are enhanced considerably when a population segment of a species exhibiting the differences becomes isolated, by some means, from gene exchange with other population segments of the species lacking the differences. Isolating mechanisms may be spatial, physiological, or behavioral. Although it may not seem that behavioral isolation applies to adaptation in plants, evidence for this hypothesis will be discussed later in relation to coevolutionary mechanisms.

The Origin of the Angiosperms   No form of life—even those forms presently extant—is a stable entity, destined to remain forever as the form now exists. Just as in the past, the influence of natural selection, acting through mutation, recombination, and genetic isolation, will continue to have a kaleidoscopic effect on the diversity of our biosphere. That natural selective factors can transform the very face of the earth is illustrated by the origin and evolution of the now-dominant angiosperm flora. Imagine, for a moment, a world with no grass, flowers, fruits, grains—a world with very few mammals and no humans. Such a world was the pre-angiosperm realm of the seed ferns, giant lycopods, cycads, and their foragers—the reptiles. The first fossils of plants suspected to be the ancestors of modern seed plants appear in early rock strata of the Triassic Period of the Mesozoic Era some 200 million years ago. These fossils become more abundant in the strata of a later period of the same Mesozoic Era, the Jurassic Period, extending from 180 to

Scale of Geologic Time

Era	Period and epoch	Plants and animals
Cenozoic began 60–70 million years ago	Quaternary Recent Pleistocene Pliocene Miocene Tertiary Oligocene Eocene (? Paleocene)	Dominance of flowering plants, mammals, and insects
Mesozoic began 200 million years ago	Cretaceous Jurassic Triassic	Rise of flowering plants and insects End of dinosaurs and pterodactyls Climax of gymnosperms Dominance of reptiles
Paleozoic began 500 million years ago	Permian Carboniferous Pennsylvanian Mississippian Devonian Silurian Ordovician Cambrian	Reduction of swamp plants Rise of reptiles Cold, dry weather Climax of nonseed land plants Dominance of amphibians Rise of sharks Warm, moist weather Rise of land plants Rise of fishes First land plants (pteridophytes) Abundance of corals Algae; fossil record meager First chordates Algae Dominance of invertebrates
Precambrian		Plants and animals were relatively simple, and the record is meager.

A fossil angiosperm, Fagopsis, a beech-like fossil of Tertiary age from Colorado.

130 million years ago. However, it is during the lower Cretaceous Period of the Mesozoic Era—130 to 110 million years ago—that fossil remains corresponding to modern angiosperm genera appear. Before the Cretaceous Period, the fossil record becomes increasingly obscure regarding the origin and interrelation of the angiosperms to other prevalent plant groups. Numerous groups that existed during the early Mesozoic Era show some affinity with the distinctive angiosperm trait of enclosing seeds in a carpel. However, the presence of other traits is necessary to qualify a plant as an angiosperm. A simple closed carpel is not enough. The stamens of angiosperm flowers have no structural counterpart among the several plant groups suggested as angiosperm ancestors. These groups are the seed ferns and the cycads. The stamen of an angiosperm is, theoretically, the equivalent of a whole seed fern leaf. Certainly, such homology or correspondence between parts may be possible. However, nothing exists in the fossil record to suggest that stamens had an origin from either seed ferns or cycads, although these two groups are most often named as the possible ancestors of the angiosperms. Indisputable angiosperm fossils from the Mesozoic Era do not help solve this problem. The missing links are yet to be found and interpreted. In addition, most of the fossil materials are casts or imprints of stems and leaves. The critical evidence of flowers and seeds is lacking. That many of these early angiosperm fossils are counterparts of existing modern genera is hypothetical at best, an assumption based on gross similarities of ancient leaf morphology to the structure of extant species. Leaf structure alone does not make an angiosperm.

From studies of the anatomy and morphology of extant angiosperm genera and families, scientists believe that the most primitive condition of the *existing* angiosperms, and presumably of Mesozoic forms as well, is found in the order Ranales. This order consists primarily of woody trees and shrubs in which the flowers are composed of perianth parts that are *numerous, equal, nonfused, spiral* in arrangement, and placed *below* the ovaries of the flower (*hypogynous*). Anatomists consider the type of carpel* characteristic of this angiosperm order to have arisen from the folding of a leaf around a developing megasporangium to form the typical structure of the angiosperms in general. Those students interested in a brief and concise account of the origin and evolution of the land flora should consult the text by Delevoryas listed at the end of this chapter.

*A single ovule-bearing leaf or modified leaflike structure.

Members of the three major
orders of insects coevolving
with flowering plants.   **Above**
A beetle, Coleoptera.   **Right** *A*
butterfly, Lepidoptera.
Below *A wasp, Hymenoptera.*

In addition to the problem of the sparsity of the fossil record, there remains the major unanswered question: What gave rise to the angiosperms' rapid and progressive dominance over other forms of plant life during the past 100 million years? Any facts shedding light on this topic lie hidden in the climate changes of the past or in a yet to be discovered fossil bed. Yet it is possible to form the following hypothesis based on existing evolutionary strategies of the angiosperms and some fossil evidence of early angiosperms and insects as well. The climate at the end of the Paleozoic Era, which directly preceded the Mesozoic, was cold and dry. These conditions favored the then predominant gymnosperm flora of coniferous trees and their relatives. In fact, these same conditions are met today in areas where gymnosperms still predominate. During the beginnings of the Mesozoic Era, widespread global climatic changes occurred. These corresponded with the advent of the angiosperms. Natural selective factors in the early physical environment of this era presumably favored the ascendancy of the angiosperms, as angiosperm physiologies were probably adapted to warmer, wetter climes.

Coadaptation

Now, consider the biological factors of the Jurassic Period of the Mesozoic Era. During this time, many of the orders of insects extant today were also evolving. We know, from natural history observations of our modern world, that intense interactions between angiosperms and insects exist today. These coadaptive interactions probably had their origins in the time span of the Jurassic Period. It is now generally accepted by evolutionary biologists that highly frequent and interdependent interactions between phylogenetically unrelated groups of organisms, such as food-providing flowers and their insect pollinators, can accelerate independent rates of adaptation and evolutionary change in these groups. This, in turn, leads to what has been called *coadaptation* or *coevolution*.

Insects and the Angiosperms. Insect orders such as Coleoptera (the beetles), Hymenoptera (bees, ants, and wasps), and Lepidoptera (moths and butterflies) interact intensively with flowering plants in the search for food in the form of pollen and nectar. This interaction gives the insect a natural selective advantage in terms of assured food and enhanced reproduction; the interaction provides the same for the plant because insect-mediated pollination is achieved. It is clear from several lines of evidence that insect-mediated pollination of angiosperms is a selective advantage. There are genera among the angiosperms which are wind, rather than

Above Wind-pollinated flowers.
Honey shows as darkened lines in
typical garden flowers. **Left** Viola.
Below Tropaeolum (pansy)

insect, pollinated. Such taxa show a lack of development of perianth parts, a lack of nectar, and an absence of attractive colors, yet produce copious pollen to accomplish fertilization. Insect pollinated flowers, by far the most numerous among the angiosperms, show the following characteristics: development of perianth parts, production of copious nectar, production of olfactory attractants to lure pollinators, development of pollinator-specific color patterns ranging from hummingbird-specific red to bee-specific ultraviolet, specifically engineered floral traps to insure pollination, and the production of plant biochemicals which are ingested and incorporated by the insect into its own sex attractant molecules. The insects have the following adaptations: specific morphological designs in body size, to permit the pollination of particular kinds of flowers; the development of chemoreception to plant-produced molecules; the metabolic ability to utilize specifically produced plant compounds and to resist the effects of toxic ones; behavioral interactions ranging from foraging patterns to pseudocopulation between insect and flower; and a host of others. With the increasing complexity of these interactions, the evolutionary strategy of survival for the two unrelated entities has become coadaptive: the species coevolve, the reproductive potential of both interactants is enhanced by the mutually profitable interaction. Coevolutionary interactions between angiosperms and animals involve hummingbirds, honeycreepers, bats, and humans. See the excellent and informative work by Meeuse at the end of this chapter.

Other Adaptive Interactions. Of course there are instances when the evolutionary strategies of a single plant and insect, bird, or mammal are not specifically intertwined. Many insects and birds retain an open option for interaction with a variety of plants and are known as generalized forager-pollinators. That is, the animal may visit different species or genera of plants without any biological obligation. Honeybees are a good example of this sort of generalized pollinator. At a given time, a hive will forage a particular plant species, and after the plant population has been exploited the insects will shift to yet another.

The reproductive fate of some plants is inextricably linked to a particular pollinator. When economically important plants requiring specific pollinators are cultivated far from the plants' natural range, pollination is a problem and can often be achieved only through the technical intervention of human beings. Such is the case with dates, figs, and many cultivated species of orchids.

Above *Hummingbird feeding on trumpet-vine flower, an example of bird pollination.* **Opposite** *Bumblebees visiting sage, an example of insect pollination (entomophily).* **Below** *Bat feeding in the flower of a sausage tree (Kigelia), an example of pollination by mammals.*

Coadaptation Biofeedback Model

Plant	Animal
Independent evolution	Independent evolution
Supplies animal with key biochemical component	Behavior or excreta aid in growth or propagation of plant
Key biochemical component attracts animal; plants with large quantities have advantage in survival	Availability of plant encourages use as source of key biochemical component
Presence of animal discourages development of alternative growth or propagation mechanisms	Continuing availability of plant discourages development of alternative sources for biochemical component
Depends totally on animal for survival	Depends totally on plant for survival

interdependent interaction

Plants and the Development of Culture

Plant	Cultural phase	Society
Natural selection	Opportunistic gathering	Omniverous habit; resulting changes in jaw and teeth
Incipient artificial selection	Purposeful cultivation	Transition to agriculture
Genetic engineering	Cultivation for survival; beginning technology	Induced dependence on cultivated foods
Total dependence on cultivation	Technological agriculture	Environmental pollution and climatic changes
Total dependence on technology	Environmental degradation	Assumed limit to technological innovation; survival dependent on balancing resources and needs

Plant Pollination   There is often an interplay of more than one attractive feature to achieve a plant's pollination and coadaptation with its pollinator. Even when olfactory cues are present, these may not be useful to the pollinator if the potential food source lies downwind. In such instances, visual cues become equally important in food location and subsequent pollination.

Development of specific visual colors can control the potential pollinators of a plant most effectively. For example, red is the color of bird-pollinated flowers and is totally ineffective in attracting bees, as these insects do not perceive reflected light in that particular spectral wave band. Nectar-feeding birds are stimulated by the color red. This can be illustrated by placing a few drops of food coloring in the sugar water of a hummingbird feeder along side another feeder in which the sugar water is clear. In the geographical ranges where nectar-feeding birds occur, there is an abundance of red to orange flowers which are conspicuously absent, quantitatively, outside those ranges. Bird-pollinated flowers also show development of a long, tubular corolla, a coadaptation with the bird's probing bill. Thus, the external color and the structure of the flower identify its probable pollinator.

Bee pollinated flowers have developed yet another visual cue in the form of "honey guides" which visually stimulate the foraging bee to seek the flower's nectar source. These honey guides for bees are most often invisible to the human observer, as the guides generally are reflective only of ultraviolet light, a stimulating color to bees but not to humans. That coadaptation between bees and such flowers has been achieved can be demonstrated by spraying paper flowers with an ultraviolet reflecting compound, such as the sun screen found in most commercial sun tan preparations. Once sprayed, such "useless" flowers are readily visited by bees.

Origins of Plant-Animal Coadaptation   That such animal-plant coadaptation has been evolving for the last hundred million years tells us little of the origins of these interactions. The fossil record shows nothing; and we are again reduced to hypothesis. Coevolutionary tactics possibly arose as a result of the interplay between two independently mutating systems—that of the plant and that of the animal. Although the mutations were independent, the natural selection operating on the interaction was not. The two biochemical systems probably became increasingly linked as a result of interdependent selection. As the interaction became truly coadaptive, favoring the reproductive potential of both plant and insect or other animal, the interaction evolved into a genetic and biochemical compromise. A hypothetical model for coevolutionary interaction in time is presented on page 58. Both interactants are initially adaptively independent. Initial opportunism leading to dependence is based on the production by the plant of a unique biochemical that is useful to the animal. In time, natural selection factors favor the reproductive potential of the animal population utilizing the plant product and developing a dependence on it, quantitatively or qualitatively. As behavior and biochemistry interact, necessary components become fixed in the genetic information of both plant and animal—leading to further genetic, biochemical, and morphological refinements. Interactions of this nature have produced plants with highly specialized biochemistry and morphology to assure pollination. Reciprocally, the pollinators are also specialized biochemically and behaviorally for detecting the food sources essential to their survival.

How does this coadaptation relate to the interactions of plants and humans? Human beings, in their interaction first with wild and then with domesticated plants, provide a superb example of coevolution at work. Much of the change in dentition and the resulting skull structure apparent in the direct ancestors of modern human beings is attributable to an increasingly omnivorous diet incorporating a variety of plant material. The population explosion associated with the neolithic transition to agriculture was, in all probability, a direct consequence of better nutrition and lower infant mortality, as well as a result of the change from mobile to sedentary lifestyles. As races of people became isolated and developed quite different diets based on available natural resources, subtle biochemical differences evolved. For example, there is a difference today between populations of Scandinavians' and Orientals' abilities to digest milk. Scandinavians possess the milk-digesting enzyme, lactase—having evolved as a population with a dairy culture. Orientals, lacking the resource of dairy products, have an exceedingly low incidence of the presence of this enzyme in their population.

It is certainly true that our survival is dependent on plants and the relationships we have established with them. This is true now probably more than ever before, in order to assure the energy-cultural-technological base we have created as a result of ancient and basic plant and human interactions. Plants also have been changed by humans, especially since our agricultural beginnings. Artificial selection has been a part of unconscious human agricultural techniques for millennia. Many major crops are now dependent on human beings for their propagation, so drastic has this artificial selection been. With increased genetic engineering and dependency of hybrids on high levels of chemicals and water for production, human engineered plants are increasingly the product of our own creation. The implications of this structured, dependent relationship are formulated in the co-adaptation biofeedback model on page 58.

Suggested Readings

Baker, H. G., and Baker, I. "Amino Acids in Nectar and Their Evolutionary Significance." *Nature* 241 (1973): 543–545.

Delevoryas, T. *Morphology and Evolution of Fossil Plants.* New York: Holt, Rinehart and Winston, 1963.

Ehrlich. P. R. and Raven, P. H. "Butterflies and Plants: A Study in Coevolution." *Evolution* 18 (1964): 586–608.

Grant, V. and Grant, K. A. *Hummingbirds and Their Flowers.* New York: Columbia University Press, 1968.

Linhart, Y. B. "Ecological and Behavioral Determinants of Pollen Dispersal in Hummingbird-Pollinated *Heliconia.*" *American Naturalist* 107 (1973): 511–523.

Meeuse, B. J. D. *The Story of Pollination.* New York: Ronald Press, 1961.

van der Pijl, L. and Dodson, C. H. *Orchid Flowers: Their Pollination and Evolution.* Miami: University of Miami Press, 1966.

Chapters Four and Five examine the conditions leading up to the earliest human agricultural endeavors, and to the subsequent development of agricultural practices throughout the world. In these chapters we consider not only the nature of our agricultural beginnings and the influence of agriculture on human societies, but also the environmental impact of these early agricultural techniques.

UNIT 2

CHAPTER

4

OUR AGRICULTURAL BEGINNINGS

A study of the origin of our domesticated plants is integral to an understanding of the origins of historical and contemporary civilizations, and vice versa. Domesticated plants provide the energy base for the growth and development of urbanized, material cultures. To understand the origins of those cultures, it is necessary to study the periods during which human beings began the transition from gathering most of their food to cultivating it.

Why plants instead of animals? In fact, the domestication of animals can sometimes provide a stable food base to support the growth of a material culture. However, historical and contemporary civilizations that have achieved notable levels of sophistication have been founded upon successful plant agriculture—probably because plants provide a food base more conserving of energy.

The Findings of Archeology and Palynology   Archeological excavations and palynology (the study of pollen) tell us, in part, how human beings first domesticated crops. The agricultural "experiment" began more than 10,000 years ago and was flourishing by 6000–3000 BC in the areas known today as Asia, Asia Minor, and Central America. Thus, the ancient writings available to us are all from periods occurring after agriculture had become an established facet of life.

It is only during the past several decades that botanists and archeologists have been able to shed new light on agricultural origins. Evidence about our agricultural past is often no more than a few charred seeds or broken animal bones extracted from the debris of long-dead civilizations, or even more primitive social groups. Such artifacts are preserved only if conditions are ideal. The best archeological information comes from dry areas where remains are better protected against the ravages of humidity, microbes and insects, and other elements causing the disintegration of plant and animal parts, pottery, fabrics, and the other items through which we reconstruct the story of the human cultural past.

These remains—the stem of a pumpkin, a few beans, chaff from grain, or the grains themselves—may be preserved in a number of ways. *Middens,* the refuse heaps found at the sites of permanent or temporary settlements, are usually valuable reservoirs of information. Agricultural remnants may be impressions, accidental or intentional designs, in ceramic wares or in the clay used in building walls. Seeds or bits of bone, charred in the ashes of some ancient cooking fire, escaping decomposition can be a rich source for the botanist and archeologist. Fragments of food material are preserved in fossilized human and animal excrement (*coprolites*) and can usually be identified by a skillful investigator. In fact, it is often possible to determine whether the food was eaten raw, or cooked. Any artifact from a culture's material inventory may offer further insight into the nature of a people's existence.

The Carbon¹⁴ Dating System The use of radioactive *isotopes*, especially ^{14}C (Carbon¹⁴) has aided researchers immensely in determining the approximate ages of plant and other organic remains. An isotope is one of any two or more forms of an element which varies only in atomic weight. All radioactive isotopes have a particular *half-life*, defined as the length of time necessary for the disintegration of half of a radioactive element's atoms. This gives investigators the key for determining the age of organic remains.

Of the 92 naturally-occurring elements, all living organisms contain these: carbon, hydrogen, nitrogen, oxygen, phosphorous, and sulfur. These six form a group indispensible to life as we know it on this planet. Carbon is abundant in all plants and animals and in the organic compounds which they produce. A radioactive isotope, ^{14}C, is the result of the impact of neutrons on carbon atoms in the biosphere. This radioactive isotope occurs in a uniform and fixed proportion to the normal isotope of ^{12}C which is present in all organisms. When a plant or animal dies, there is no more addition of ^{14}C into its system. The ^{14}C isotopes already in the organism's system proceed to disintegrate at a rate indicated by the half-life of 5560 years. The quantity of ^{14}C in living organisms is not great to begin with; it is therefore difficult to measure it in materials greater than 40,000 years of age. However, it is a valuable tool in analyzing remains found in the settlements of incipiently agricultural peoples and, in particular, those organic remains directly associated with the practice agriculture. ^{14}C has been used extensively in dating archeological specimens of plants, human bones, teeth, and animal remains from the period that concerns us—from 10,000 to 1000 BC—in both New and Old World investigations.

The Invention of Tools What kinds of evidence do we have concerning human activities in the period leading up to the agricultural transition which occurred independently in various areas of the world between 15,000 and 5000 years ago? The invention of tools is a prerequisite for all but the most primitive agriculture, but tools themselves do not insure the innovation of agriculture. Tool-making, in sophisticated forms, was present in the cultures of *Homo erectus, H. habilis,* and *H. neanderthalensis* long before agriculture was practiced. By

250,000 years ago, hominids lived in many different areas and had adopted standard tools and techniques for making them, although populations were widely separated. Such a homogeneity among the tools of early human cultures suggests that these various populations experienced similar environmental challenges and adopted common strategies for dealing with them. By 75,000 years ago, the tools of these wide-ranging groups had become sufficiently specialized to indicate to investigators that the tools were adapted for local food-gathering operations. By 50,000 years ago, a new set of tools is present in the record. These correspond with the appearance of contemporary humans, *Homo sapiens*.

When Did the Transition Occur?   Traditionally, archeologists have referred to three stages in the development of human culture: the Paleolithic (Old Stone Age); the Mesolithic (Middle Stone Age); the Neolithic (New Stone Age). These phases of advancement are characteristic of Europe and the Near and Middle East. However, recent investigations do not bear out that these periods are universal in human history. Accordingly, we avoid reference to them except in relation to those areas where the three periods are specifically valid. The incorporation of agriculture into man's culture is frequently referred to as "a Neolithic event." In fact, the adoption of agriculture on a worldwide basis occurred at different times in different areas (see Chapter 5).

The archeological record suggests that plant domestication in both the New and Old World was initiated primarily in dry, hilly, or mountainous regions. But as the record itself is so much more completely preserved in these less humid terrains, investigators may have gotten a lopsided perspective. In drawing assumptions from available data, it is important to realize that archeology can offer only a fragmentary view of human history, one subject to misinterpretation and other human error.

References to the innovation of agriculture as a "revolution" may be misleading, conveying the impression that plant cultivation was adopted suddenly by human beings. In reality, agricultural development must have been a long process—one of trial, error, accident and invention. Nor would it be correct to assume the origins of plant or animal agriculture were the same from one region to another. Climate, available resources, and random chance must all have been determining factors.

From Hunting and Gathering to Agriculture   How does a group of human beings make the transition from hunting and gathering to a more sedentary agricultural lifestyle? The history of a group of North American pre-Columbian Indians provides insight into the process. Coe and Flannery (1964) suggest that *evolution* is a more appropriate term than *revolution* for the change to agricultural lifestyle. Shoshonean bands occupied the Great Basin of the United States until the mid-19th century. They had a lifestyle which was probably quite similar to that of the pre-agricultural Mesoamerican Indians of the fifth millennium BC, who were the first peoples, as far as we know, to begin domesticating maize.

The nature of the Great Basin is essentially desert. The lack of irrigation techniques hindered the evolution of stable agriculture. Some bands did sow wild grasses, and at least one attempted an ineffectual watering of wild crops. However, the Great Basin Indians were dependent primarily upon hunting and gathering for their sustenance, a condition that kept population densities low and material culture simple.

It is a fact that most primitive human cultures, although limited by the character of the *macroenvironment* (in this case, the Great Basin desert), will take advantage of numerous *microenvironments*, varying niches within the broader environmental pattern.

The Great Basin Indians did this and took advantage of microenvironments within the desert (Steward, 1938)—moving on a predictable seasonal round among vertically and horizontally distinguishable microenvironments, from low salt flats to higher piñon pine forests. Although the total aspect of the desert circumscribed the potential of their existence, they *chose* the microenvironments, or niches, vital to them. These microenvironments affected many aspects of their culture—the limited technology, social patterns, and settlement patterns. Among any pre-agricultural peoples who finally adopted agriculture, there must have been a similar, systematic exploitation of the environment based on more than mere chance.

The First Domestication of Plants   The domestication of plants probably began in a very peripheral fashion, perhaps with the planting of a few desired varieties, while hunting and gathering remained by far the most important means of obtaining food. Then, gradually, the convenience and predictability of cultivated plants assumed an increasing importance. The nature of the environment would be a significant factor in this process.

In what environment did agriculture develop first? One might expect the answer to be—in lush, fertile lands. If, in fact, the archeological indication is correct, then agriculture tended to develop in more arid places—usually less bountiful

Above *A farm granary for the storage of maize ears in Tanzania. Improved methods of pest control and better storage facilities for agricultural produce are two interrelated ways of cutting down on waste and increasing income from agriculture.*

Left *In Chad, farmers build grain stores of dried mud and straw. (FAO photos)*

regions. A rational explanation of this phenomenon is that human beings had more of an impetus to experiment with the means for insuring a better supply of food when food was not as abundant. For example, the Indians of the northwest coast of North America had an elaborate culture, yet the cultivation of crops never became part of their way of life because it was not necessary. In the relatively mild, damp climate of that area, food was plentiful in the form of numerous terrestrial animals, salmon and other fish, berries, and plants.

It seems reasonable to assume that the development of agriculture was stimulated by need, rather than by a desire to grow plants or by the realization that efficient agriculture provides food surpluses and a greater potential for human society. As the latter is a product of many centuries of agricultural experimentation, the concept requires the perspective of time. Early inhabitants of coastal Peru appear to have abandoned their farming whenever fish became plentiful (Lanning 1965).

What kind of terrain was most conducive to the initial cultivation of plants? Investigators disagree on the answer. It may have been grasslands, but it can be argued that the felling of forest is easier than the removal of sod. Probably, choice of terrain varied according to local circumstances and different inclinations. Anderson (1952) has suggested that the first farmland may have been refuse-collection areas. These would have been relatively free of other vegetation, yet probably more fertile due to the accumulation of organic debris.

On what criteria were plants selected for cultivation? Not having any specific historical records, we can only deduce the answer according to logical conclusions. Again, there is no universal agreement among investigators. One theory proposes that the plants first cultivated were common kinds already being used as food. Others argue that there would be no reason to cultivate plants already abundant; therefore the first crops were those useful varieties less easily obtained in nature.

The former perspective probably has greater validity. The beginnings of agriculture surely involved casual processes, such as the planting of individuals of an existing food source after noticing that they had grown in the settlement compound from seeds tossed away among rubbish. It would be a dubious assumption to imagine that pre-agricultural hunters and gatherers, during the initial stage of plant cultivation, decided consciously that a particular plant would serve them as a crop. Rather, intentional planting came about, in all probability, only after the concept of doing so had been grasped via indirect means.

The Earliest Cultigens   Certain plants must also have been more adaptable to systematic agriculture than others; or, in the words of Paul Mangelsdorf (1952), "more mutable." He postulates that the ancestors of cultivated plants were not well adapted to living in the wild—underscoring the fact that many domesticated plants are very poorly represented in nature. Certainly a greater mutability is an advantage in cultivating a plant. It is easier to breed for desirable qualities, such as larger, juicier fruits, or for other agriculturally important characteristics which might be irrelevant to the plant's success in a wild state.

A further element that may have facilitated the cultivation of a given plant is the isolation of the *cultigen* (the cultivated plant) from other individuals of the same species growing in the wild. When agriculture began to be an incorporated aspect of cultures, certain plants may have been removed from their natural habitats—wooded hillsides, for example—to be grown on flat, open country. If they survived at all, the change of environments would have been enough to cause genetic isolation, even though the actual distance separating the crop from the wild plants may not have been great. This human-induced separation of the crop plant from wild ones would have made it easier to select for desired characteristics.

The length of time involved in completely domesticating a crop plant would have varied according to the species, the persistence in cultivating it, and many other factors. Research on the length of time required for the domestication of wheat and barley in Asia indicates that the process spanned about 1500 years (Helbaek 1966). With other plants, considerably less time may have been involved.

The shift from depending on wild food to more reliance on the yields of domesticated plants and animals caused profound changes in human culture. For hundreds of thousands of years, people had lived in relatively small, scattered bands. The energy base of the band was dependent upon the products of hunting and gathering. There were surely times when these early human beings experienced abundance, and times when food shortages caused starvation and deaths. What is significant is that, whether in times of plenty or in times of deprivation, these people essentially had no control over their situation. They could not increase the quantity of game and wild plants, or alter the nature of them.

Agriculture and Lifestyle Changes   As long as hunting and gathering formed the basis for human existence, human beings were limited in their numbers. However, limited food resources may not have been the reason. According to the demographer Edward Deevey, the way of life, rather than the availability of food, contained the human population (Deevey, 1960). Hunters and gatherers were mobile peoples. They could not support numerous children, and, in societies in which adults lived to be only 25 or at the most 30 years old it would have required numerous offspring to accelerate population growth.

It seems probable that the above explanation and the availability of food resources were both factors in restricting the population increase in hunting and gathering societies. What is unequivocally true is that, whatever the actual size of the population, the ramifications of advanced civilizations could not have been developed without the *food surpluses* provided by agriculture. Surpluses are basic to the maintenance of any sophisticated culture, including contemporary ones, in which a large percentage of the population is engaged in activities other than food-getting. For a society to support its artisans, craftsmen, and others specializing in occupations unrelated to the procurement of food, there must be a means of supplying food in a multiple of the quantity needed by the cultivator's immediate family. In addition, if the society is to have the stability for continued specialization and sophistication, there must be insurances against times of scarcity—a security that hunters and gatherers could not provide.

As plant agriculture assumed increasing importance in early human societies, lifestyles gradually altered. People turned from the mobility associated with hunting and gathering to a semi- or totally sedentary village life. The permanence of village life and the security of a more predictable food source opened avenues for further cultural specialization. Among pre-agricultural peoples, there were finely-made tools; but these were related to food procurement. Probably the hunters were expected to be able to fashion their own equipment. In sedentary, agricultural societies, individual specialization became possible. Some people were shamans and others makers of pottery, for example. These distinctions were probably not rigid. A person adept at producing ceramic ware may still have had to cultivate his or her own crops. However, in the free time and relative security offered by the practice of agriculture, we find the beginnings of cultural specialization. For the first time, an individual's means of support was not always related directly to food production. Clearly, without agriculture human societies would have remained at a much more basic level.

Suggested Readings

Anderson, Edgar. *Plants, Man, and Life.* Boston: Little, Brown and Co., 1952.

Coe, Michael D., and Flannery, Kent V. "Microenvironments and Mesoamerican Prehistory." *Science* 143 (1964): 650–654.

Deevey, Edward S., Jr. "The Human Population." *Scientific American* 203 (1960): 194–204.

Heiser, Charles B., Jr. "Some Considerations of Early Plant Domestication." *Bioscience* 19 (1970): 228–231.

Helbaek, Hans. "Commentary On The Phylogenesis of *Triticum* and *Hordeum.*" *Economic Botany* 20 (1966): 350–360.

Lanning, Edward. "Early Man In Peru." *Scientific American* 213 (1965): 68–76.

Mangelsdorf, Paul C. "Evolution Under Domestication." *American Naturalist* 86 (1952): 65–77.

Steward, H. J. "Basin-Plateau Aboriginal Socio-political Groups." *Smithsonian Institution Bur. American Ethnological Bulletin* 120 (1938).

CHAPTER

5

HISTORICAL AGRICULTURE IN THE NEAR EAST, ASIA, AND THE NEW WORLD

After the development of tool-using, the second major phase of the human evolution into cultural beings came with the domestication of plants and animals. This resulted in the transition from hunting and gathering to a semi-agricultural lifestyle which, in most populations, finally culminated in sedentary, urban societies. These societies were, of course, dependent upon the cultivation of food materials.

Agriculture arose independently in at least several different areas of the world. Archeological data does not give us a complete record of the events comprising the "agricultural revolution." In addition to evidence from the past, we can also gather some information about these first cultivators by observing small populations of people still living today at a cultural level equivalent to that existing among human beings at the dawn of agriculture. One of these populations is the Kalahari Bushmen, who engage in food collecting with little attempt to control or stabilize the food production effort.

The Middle East   One of the most intensively studied sites of early agricultural development in the Middle East is Jarmo, in the Kurdistan region of what is currently Iraq. Excavations conducted at Jarmo under the direction of Robert Braidwood (Braidwood, 1960) have given us insight into the pattern of agricultural growth in this area.

It appears that bands of people moved from caves to the open meadows of the highlands. Flint sickle blades, probably used in the cutting of wild or semidomestic grains, and milling and grinding stones were found at the site. Although the Jarmo area is exceedingly dry today, during its habitation in past millenia it was on the edge of the so-called "fertile crescent," and was more conducive to the rise of agriculture than it would be today. Investigators once assumed Jarmo to be the earliest known agricultural community. However, there is new evidence, discussed later in this chapter, that agriculture was begun in southeast Asia at least 5000 years before Jarmo.

Early Farming Villages   The two villages that have been excavated in the Middle East are Jarmo in Iraq and Tepe Sarab in Iran, both dating from 9000 to 8500 BC. Jarmo was a small but permanent year-round village of about 24 adobe houses which show evidence of repeated repair and rebuilding. Investigators estimate that approximately 150 people lived there at a density of about 27 people per square mile. This may seem a tiny settlement to us today, but 10,000 years ago there were probably few larger bands of Homo sapiens.

We know that the cultivation of food had begun at Jarmo.

Two varieties of domesticated wheat, two-rowed barley (modern barley is six-rowed), and the bones of domesticated goats and dogs were found there. At first there was probably a period of incipient domestications, parallelled by hunting and gathering, that lasted at least several hundred years.

By about 5000 BC, agricultural practice had spread from the highlands, where grains were first domesticated, into the fertile valleys of Mesopotamia. In the valleys periodic flooding and small-scale irrigation heightened potential agricultural productivity. In that period of 3000 to 4000 years, human existence changed more than it had in the preceding 500,000. For this reason, the establishment of agriculture is called a "revolution," even though the process was gradual in terms of our own time reference.

With a dependable energy base, human potential was now released for the innovation of a variety of new pursuits not directly related to procuring food. Inventions basic to modern cultures, such as weaving, pottery, plows, wheels, and metallurgy begin to appear in the archeological record. Large numbers of clay figurines representing animals and pregnant women (fertility symbols) indicate the development of a non-materialistic dimension of life. It should not be assumed, however, that pre-agricultural peoples did not have intangible religious and spiritual beliefs of considerable complexity. The agricultural revolution and the development of such crafts as ceramic sculpting expanded the means of expression; systematic cultivation provided the time for doing so.

The First Domesticated Plants How long it takes for a plant to become domesticated cannot be precisely determined. We do know that critical changes often take place and are the results of both intentional and unintentional processes of artificial selection practiced by cultivators. These early farmers probably bred the kinds of crops that could be harvested and utilized easily.

Consider the examples of wheat, a staple of life at Jarmo. Wild grasses, of which wheat is one, scatter their seeds upon maturity. In nature, this dispersal is necessary for their biological survival. Consequently, wild wheats fall apart readily and cannot be harvested easily, as most of the seeds would have to be gleaned from the ground. Yet in every population of wheat there occur a few individual plants possessing a gene that endows them with tough, resistant grain spikes which do not break readily. These plants do not disperse their seed at all well in nature, but they are ideal for human harvesting. Early farmers would have necessarily collected more

In many areas of the world
wheat is still harvested by
hand methods, just as it was
harvested by Neolithic farmers.
(FAO photo)

from the tough-headed spikes, the numbers of which then increased with each successive cycle of planting and harvesting. In the process, the tough-headed specimens of the original wild population became the domesticated population, dependent upon human cultivation propagation, since few of the seeds would become dispersed without human intervention. As we will describe later, an even more drastic accommodation to domestication was made in the case of corn.

Early Agriculture in Europe Investigators agree that the practice of agriculture and the consequent cultural changes diffused from a center of common origin in the Middle East to parts of Europe and areas farther east. However, were the entire historical record of agriculture available, the pattern of agricultural spread would not be so simple. Certainly there was a gradual cultural diffusion from Middle Eastern regions to both the East and the West.

How was agriculture established in Europe? One theory is that Middle Eastern peoples migrated there, bringing their cultural knowledge. This seems likely because the first archeological record of farming in Europe shows a very advanced tool technology. The best explanation for such tools, without a prior record of simpler implements, is the sudden arrival of new inhabitants who brought their already perfected tools with them. Some investigators suggest that this invasion was by enclaves of *Homo sapiens* from Anatolia in the Near East, and that they supplanted *Homo neanderthalensis*, who was at the time the sole *hominid* (human-like) resident of Europe. Whether this was accomplished through struggle or by the superior abilities of *H. sapiens* in competing for resources, or both, is not known. At any rate, *H. neanderthalensis* became extinct and was replaced by the aggressive populations of *H. sapiens.*

An Anthropologist's Experiment in Denmark Research conducted by Danish anthropologists has added a great deal to our understanding of the beginnings of agriculture in Europe. Much of the data has been obtained from comparative pollen analysis. Pollen is particularly suited to preservation in bog and lake sediments due to the extremely hard and microorganismically indigestible *exine*, or outer cell wall layer, of the pollen grain. The quantity and specific kinds of pollen in a sample can give considerable information about the historical vegetation of an area.

From these pollen profiles, botanists can tell the species makeup of the historical plant communities. If species are present that would not be a component of the naturally structured biological community, the probability is high that there was some kind of disruption. This might be a natural event, such as fire, or the intentional clearing and farming of an area. Sometimes pollen samples from agriculturally important crops are the only evidence of agricultural manipulation in an area.

Denmark is an especially profitable region for archeological investigation of so-called Mesolithic and Neolithic sites because of the presence of numerous heath areas and bogs which contain high levels of tannic acid. This chemical retards bacteria and fungal decay; and, consequently, aids in preserving pollen, plant fragments, and wooden artifacts in the bogs. From the investigations of pollen profiles from such areas, the following story of agricultural growth was unravelled (Iverson 1956).

Toward the end of the last Ice Age, as the ice sheets were receding and vegetation was returning, bands of hunters ranged over Denmark. As forest area expanded, resulting in a reduction of larger game animals on which the people had previously depended, the human beings retreated to coastal areas. There they became fishermen who practiced agriculture to a very limited extent. This lifestyle apparently continued for some time, until the beginning of what is termed the Neolithic period, when human groups began to establish themselves in forested regions. Clearings were hewn, as evidenced by the diminishing pollen record for trees. At the same time, there is the sudden appearance of the pollen of herbaceous plants—a further indication of clearing —and the pollen of cereal grains. Thereafter, the pollen profile indicates a proliferation of trees which usually grow after forest clearance—birch, willow, and aspen. In the third and final stage, the birches and grasses declined and were replaced by larger trees. According to the pollen record, some of the settlements lasted for less than 50 years. Presumably, these early peoples used the cleared and burned areas first for planting cereals; then for pasture; and finally moved to other areas, allowing forest to regenerate.

No one had offered any data-supported hypotheses of how Neolithic farming was accomplished, until Iverson's research team carried out an experiment in Neolithic farming methods. They wanted to prove two points: first, that Neolithic peoples were able to clear forests with the tools at their disposal, and second, that fire clearing was an essential part of their agricultural practices. A practical application was deemed necessary to support the contentions of these hypotheses.

Iverson and his coworkers cleared a 2-acre area in the Dravin forest of Denmark, a space similar to those cultivated by the early farmers. They used Neolithic flint blades from the Danish National Museum. Wooden *hafts,* or handles for the blades were replicated from ones discovered in a Neolithic bog deposit. The workers had little difficulty in learning to use the Neolithic tools. On mastering the techniques, they were able to fell trees of more than a foot in diameter in less than half an hour, thereby negating some investigators' arguments that the primitive tools could not have been effective in forest clearance.

The next problem was learning how to burn off clearings. In the case of larger trees, it was necessary to let the trunks lie for a year before burning. It is entirely probable that crops were planted among the larger trunks—a practice common with slash-and-burn agriculture in the tropics. Immediately after burning, the plots were sown with barley and two varieties of primitive wheat. For a comparison of the impact of burning on plant growth, grains were also planted in a cleared but unburned plot. Wheat and barley grew well in the burned section, but scarcely at all on land that had been only cleared. This was probably due to the reduced acidity of the forest soil brought about by the presence of alkaline ash.

On maturity, the wheat and barley crops were harvested with flint sickle blades. Vegetation succession patterns were observed on both the burned and unburned plots. It was found that the burned plots' vegetation records matched those suggested by the pollen profiles of the historic sites.

The above experiment provides insight into the methods used by early farmers in Europe, although specifics no doubt varied according to different environmental demands. Conditions at Jarmo, for example, would not have presented the same problems as those in a European forest. However, there was undoubtedly diffusion of culture (and therefore of agricultural practices) from the Middle East into the Near East and Europe. In different areas of the world, agriculture arose independently and not necessarily by identical steps.

The Origins of Agriculture in Southeast Asia Until recently, it was believed that there were two areas in which agriculture was innovated and from which it spread to other areas. These were the Middle East and the central part of what is now Mexico. Yet there is now evidence that a separate agricultural "revolution" occurred in Southeast Asia perhaps 5000 or more years before the practice of agriculture at any other known sites (Solheim 1972). It seems obvious that this was an independent

Rice terraces in western Java. (FAO photo)

cultural phenomenon since the plants and animals involved were largely unknown in the Middle East.

Archeological investigations conducted by William Solheim and his colleagues in Thailand hint at the surprising conclusion that during the period from 13000 BC to 4000 BC some of the most technologically sophisticated civilizations in the world flourished on the mainland of northern Southeast Asia. In some aspects, these civilizations probably surpassed those later ones in the Middle East and the Mediterranean. These discoveries not only question former assumptions, but point out that our historical record is incomplete.

Two Key Archeological Sites   The two sites from which the evidence was obtained are a prehistoric mound, Non Nok Tha, close to the area that was flooded by the waters of the Nam Phong Reservoir in northern Thailand, and Spirit Cave, in the northwestern corner of Cambodia. On its discovery, Non Nok Tha was being tilled jointly by a few farmers from a small hamlet. It was mostly covered in plantings of bananas, peppers, and mulberry bushes; but two areas were cleared for the threshing of rice from nearby paddies at harvest time.

On excavating a section of the mound, it was found that the upper strata contained a few iron implements. Lower levels yielded earlier graves of people who had not been cremated, along with a variety of grave furnishings—including stone molds used in the casting of bronze age blades, a few of the axes themselves, miscellaneous other items of bronze, and polished stone tools. There was pottery, well-made but with few decorations. In the lowest levels, there were more stone tools and some decorated pottery, but no bronze objects of any kind. These finds were significant for several reasons. This was the first site in Southeast Asia showing a long interval when bronze was known at a time when iron was not. Also, the pottery was decorated with patterns characteristic of samples from the Philippines and other areas in Southeast Asia, suggesting cultural contacts.

From an agricultural standpoint, Non Nok Tha's offerings were stimulating. Some of the pottery from the lowest levels bore the imprints of cereal grains and husks. Samples sent to the Kihara Institute for Biological Research in Japan confirmed the imprints as *Oryza sativa*, the species of rice most abundantly grown in Asia today. Animal bones were also present, positioned as though they had been severed from large animals and placed in the graves, perhaps as funeral offerings. Upon examination, they proved to be indistinguishable from bones of the modern *Bos indicus*, the common humped cattle of India. The oldest levels at Non Nok Tha were radiocarbon-dated as 5000 to 6000 years old; presumably, contemporary rice and cattle were already well established by that time.

Spirit cave, excavated by Solheim's student Chester Gorman, held even more intriguing finds. Opening high on a limestone cliff overlooking a stream near the border of Thailand and Cambodia, the earliest strata were probably formed about 10,000 BC. Its last human occupation occurred approximately 5600 BC. Artifacts from all levels, except the final 1200 years, were simple stone tools apparently representative of the Hoabinhians, who were hunters and gatherers of southeast Asia.

Upon superficial examination, Spirit Cave showed signs of being just one more Hoabinhian site, of interest only because it is farther west than other such sites. However, two findings soon attracted special attention. First, sifting of the soil produced plant remains of 10 separate genera—including pepper, butternut, almond, candlenut, and betel nut. Second, and even more important were the remains of a species of cucumber, a bottle gourd, Chinese water chestnuts, and some legumes: peas (*Pisum*), beans (either *Phaseolus* or *Vicia*), and possibly soybeans (*Glycine*). Although conclusive proof is lacking, it is probable that at least some of these were cultivated. Even if the plants were collected from the countryside, they represent a stage of hunting and gathering at least as advanced as that existing in the Middle East during the same period. However, if any of the plants were being cultivated, then the inhabitants of Spirit Cave were engaging in horticulture for a minimum of 2000 years before the proposed date of the first plant domestication in the Middle East.

Another major finding at Spirit Cave was the evidence that a clearly non-Hoabinhian culture had begun occupation by no later than 6800 BC. This was evident from distinctive artifacts, such as rectangular stone adzes with partially polished surfaces, and knives fashioned from grinding both sides of flat pieces of slate, which appear at that time point in the excavation strata. Tools of this kind have not been uncovered from such early levels at any other Southeast Asian sites.

In the upper strata, there were potsherds (pieces of pottery) showing variety in finish and design. Except for a few samples unearthed in Japan, the Spirit Cave potsherds are the oldest in the world. Their variation in design and finishes indicates that a pottery tradition had already existed for some time. In the evolution of pottery, there is characteristically a period, of indeterminate length, during which the ceramics are simple and undecorated. Gradually, the craft development progresses toward more complicated forms and patterns. If adorned ceramics are present in an archeological sample, one can be certain that there is a ceramic tradition with a significant precedence.

Prehistory Time Periods in Southeast Asia   As we have mentioned, the conventional terminology of prehistory is separated into the Paleolithic, Mesolithic, and Neolithic periods. Agriculture did not begin until the latest of these. This arbitrary classification of periods is based on the history of the human race in Europe and the Middle East; it is not necessarily applicable to other areas. For Southeast Asia, Solheim has suggested, tentatively, new terms for the stages of cultural evolution. In the *Lithic* Period, humans used chipped and flaked stones for tools; this period lasted until $\pm$ 40,000 BC. The *Lignic* Period (derived from the Latin *lignus,* wood) was characterized by an increasing use of wooden implements, especially bamboo, which has continued in importance until the present. This period ended about 22,000 BC, when the Lignic gave way to the *Crystallic* Period, during which definitive cultures were formed in Southeast Asia. It was during this period that the domestication of plants was initiated and established—at least in some areas—as a means of stable food procurement by 13,000 BC. The culture represented in the lowest levels of Spirit Cave was probably one in which horticulture was evolving into a generalized domestication of plants and animals. No doubt different plants were selected from one region to another, and as time passed they were culturally diffused. The same may be assumed for domesticated animals: pigs, chickens, cattle, and perhaps dogs.

The fourth period suggested by Solheim is the *Extensionistic* (10,000–2000 BC), in which people began to move out of the mountains and into other areas of Southeast Asia—taking with them their domesticated plants and animals. The transition to life in the lower piedmont adjacent to the mountains must have facilitated agricultural development, for the terrain would have been more favorable. Rice may have been one of the plants transported from the highlands by the inhabitants of Spirit Cave. The slate knives found in upper levels of the site closely resemble those still used to harvest rice in parts of Indonesia.

Agriculture and Cultural Evolution in Southeast Asia   It is interesting that the cultural evolution in Southeast Asia did not parallel that of the Middle East. There was no rise of cities or centralization of political power. Even as recently as 2000 and 3000 years ago, fortifications are absent from any part of that Southeast Asian region, an indication that organized warfare did not occur or that it followed different rules. With one exception, the scattered cultures of Southeast Asia seem to have had contact with one other, while retaining political autonomy. The exception is the area comprising what is currently North Vietnam and bordering on parts of southern China. Here, there seems to have been a central authority independent of the imperialistic dynasties of northern China. Further investigations, however, are needed to confirm the existence of this supposed state.

Southeast Asia's history refutes the theory that warfare is always an outcome of the sedentary, geographically stabilized life that evolves with the growth of and dependence on agriculture. The lack of large cities and central governments were undoubtedly factors in keeping existence relatively peaceful in Asia during that period. Why did large cities fail to develop during so long a period of established agriculture? Investigators have not found the answer. However, the fact itself is also proof that the practice of agriculture as part of a cultural fabric does not automatically lead to the development of sophisticated urban centers. It also serves as an example of the logical fallacies that arise in science, or in any discipline, from extracting generalized notions from particular instances. In the Middle East, Europe, and the New World, sophisticated material cultures followed the advent of agriculture within a relatively short time. But such a course is not inevitable; it is merely one tactic among various possibilities.

It was not until 2000 years ago that the first centralized states were becoming established in Southeast Asia. Solheim calls this era *The Period of Conflicting Empires*. Even then, the change was not generated from within the existing cultures, but came about in response to the political and religious influence of India. A number of petty states existed in Southeast Asia for about 1500 years. These were supplanted during the remaining 500 by European influence, a change which was largely irrelevant to the majority of native inhabitants. This period ended with the termination of World War II. Then came a time of flux, which was marked by warfare, the disintegration of traditional boundaries, and the death of philosophies that had survived for millenia. Even now, primitive subsistence agriculture provides a way of life for most of the peasants of war-plundered Southeast Asia.

These recent findings in the archeological record of Southeast Asia's agricultural history are important precisely because the facts deviate from traditional academic expectations. In the area of the world where the human experience with agriculture may have been the longest, there were, until outside influences intervened, no complex city-states devoting their energies to warfare and imperialistic conquest. The way of life remained rural, materially simple (although some tools were quite sophisticated), and socially peaceful. Why this was so is open to speculation. In any event, the Spirit Cave site gives us another perspective on the relationship between agriculture and cultural development.

New World Agriculture It was long thought that the cultures of the New World were of a much later origin and considerably less complex in their development than those of the Middle East and Europe. However, during the past 30 years extensive investigation by anthropologists and botanists has established that human beings have been permanent residents of the American continents for at least 30,000 years, and perhaps much longer.

During the period from 1948–1960, Paul Mangelsdorf and Richard MacNeish and others made numerous discoveries relating to early human habitation, lifestyle, and incipient agriculture in Mexico. One of the more striking aspects of their data revealed differences in the ways New World people achieved partial and total control over their resources, in contrast with their counterparts in the Old World.*

In the Old World, Mesolithic and Neolithic peoples domesticated a variety of animals, among them horses, cattle, sheep, goats, rabbits, certain domestic fowl, and dogs. These livestock furnished raw materials for many cultural basics: meat, milk, hides, wool, and as beasts of burden. In the New World, there were fewer domesticated animals, and many of these were half-wild. In the Andes, domestic animals included only a group of cameloids—llamas, alpacas, and vicuñas—and the guinea pig. In Central and North America, dogs and turkeys were the only really domesticated creatures. Both New and Old Worlds had refined the practice of beekeeping.

New World peoples, especially those of Central and South America, were faced with less potentially domesticable animals. They depended on hunting to supply them with animal proteins. This lack of animal resources, coupled with the advantage of a rich tropical and subtropical flora, is reflected

*We will use the designation "Old World" for Europe and the Near and Middle East. As we pointed out, Southeast Asia had a unique course of agricultural evolution.

Some varieties of corn native to the New World. These varieties
are preserved in seed banks to insure the genetic variation of the
genus Zea, and may be used in subsequent hybridization
experiments. (USDA photo)

in the great number of plants domesticated by New World peoples. In contrast with the drier Old World regions where civilizations arose (with the exception of India), the New World areas were much more lush. In Europe and the Near and Middle East there was a relatively small number of domesticated plants—wheat, millet, rye, barley, cotton, figs, citrus, and seasonal herbs used as vegetables. The number of domesticated flora in the New World was huge by comparison (see box). Even so, New World civilizations were *based* upon a small number of staples.

Maize: The New World Staple   Of the many domesticated plants of the New World, *maize*—or Indian corn—has commanded the most attention. This crop was the energy base for the generation of civilizations from the St. Lawrence River on the North American continent to Chile in South America. At the time of the European conquest of the New World, there were more than 150 varieties of corn in cultivation. Because it grew throughout the New World and was of tremendous cultural importance, investigators assumed that, could the place of origin be found for corn, there, too, might be found the site where agriculture and civilization originated in the New World.

Due principally to the efforts of Paul Mangelsdorf and Richard MacNeisch, it is now possible to date one archeological site in Mexico, in the region of Tehuacan, as being agricultural and dating from 7000 to 3000 BC. Mangelsdorf also contends that earlier speculations designating *teosinte* grass or *Tripsacum* as the ancestral plants of corn are wrong. Corn is descended merely from an ancestral corn (Mangelsdorf, et al. 1964). Although a wild corn variety had never been found, he constructed a model of the assumed ancestor, which looks very similar to a small popcorn having kernels each loosely encased in a sheath. Subsequent investigations support this theory. However, there are also persuasive reasons for accepting the earlier theory (see Chapter 6). Pollen thought to be from wild corn was discovered in the lake bed of Texcoco on which Mexico City now stands. The cores from which the samples were obtained are at least 80,000 years old, antedating by some 50,000 years the arrival of human beings in the New World. Whether or not the pollen is, in fact, from wild corn is questionable. Nevertheless, Mangelsdorf's studies have shed light on the growth of agriculture in the New World.

In 1948, Herbert Dick explored a cave in New Mexico. At the lowest levels of excavation he found some tiny corn cobs which were dated by the ^{14}C method of 3000–2000 BC. Even these primitive samples of corn did not fit the model Mangelsdorf had designed, and clearly were still a domesticated variety of corn. Workers then turned farther south in attempts to locate older sites which might yield the true wild progenitor of modern corn. Excavations in Guatemala and Honduras resulted in the finding of more small cobs of approximately the same age as those uncovered in New Mexico—a clue that there had been cultural diffusion from some point midway between those two areas. The presumption was only a guess, as corn might have come from some point farther south, or even farther north. But having no more definitive information, workers oriented themselves toward Mexico.

Chiapas, Mexico's southernmost state on the Pacific Coast, yielded nothing of importance in the search for corn's ancestry. Needing dryness for the development of its seeds, corn could not have tolerated the humid forests of that area. The highlands were a possibility, yet they still were more humid than locations farther north. So researchers commenced excavations just south of Puebla located in the Tehuacan Valley, which has a relatively dry terrain. In 1960 preliminary efforts uncovered cobs of corn dated at 5600 years, the oldest yet found. This was also a domesticated variety; the wild ancestor still remained elusive.

Finally a cave at Coxcatlan (near Tehuacan) was excavated and shown to have 28 separate levels of human occupation, the earliest of which dated back 10,000 years. The site is unique. It is remarkable for its numerous evidences of human habitation, in the form of animal bones—including extinct Pleistocene species—and fossil plant materials. Also, it has been occupied almost continuously, a pattern unlike any Old World excavations. At last, miniature ears of corn having few kernels, a reduced cob, and a papery sheath encasing the kernels were discovered. As dated by ^{14}C, the corn was about 7000 years old.

Whatever the specific origin of corn, it is certain that the people of the Tehuacan Valley were among the first in the New World to undergo the transition from a primitive hunting and gathering means of subsistence to agriculture. The process has been traced, with intriguing detail, for a period of more than 9000 years. The archeological record pieced together from the excavations at Coxcatlan provides us with a relatively clear picture of the development of agriculture in Mesoamerica.

The Tehuacan Valley: Site of Early New World Agriculture   From 12,000 to 9000 years ago, the archeological record shows that there were few people in the valley, probably a small group that hunted wild game and gathered a variety of edible plants. Bones from a Pleistocene horse and other animals were found in the lowest levels of the cave, along with a few rudimentary tools. Approximately 8700 years ago, hunting techniques changed, and more specialized tools appear. By 5000 years BC some plant cultivation had begun; investigators found the remains of domestic squashes. Grinders, *metates* (stone grinding bowls), and pestles of stone were also present. At this time, however, only an estimated 10% of food was agriculturally produced; the remainder still came from hunting and gathering.

During the following two millennia, an inventory of domesticated plants appears: amaranth, bottle gourds, squashes, zapotes, beans, chilies, and corn. A fundamental shift occurred. By 3400 BC, 30% of the food was a result of agricultural production. Domesticated turkeys and dogs were established, and the construction of pit houses in the floor of the cave had commenced. A thousand years later, there were hybridized forms of corn, and ceramics. A proliferation of the material culture and a steep rise in the population density occurred at the same time. The nonmaterial culture also became more complex. An extensive figurine cult appeared, along with notable refinements in pottery, and ceremonial burials.

By 850 BC, the people of Tehuacan had begun growing corn in irrigated fields in the valley. There is evidence of commerce with other cultures, and the construction of temple mounds. They probably had some contact with the little-known Olmecs of Veracruz, who left a legacy of huge, monolithic stone heads. But within 600 years, influence seems to have shifted from the Olmecs to the Monte Alban culture in Oaxaca. New domesticated plants, peanuts and tomatoes, were introduced during this period.

By 700 AD, influence from the Mixtecs, centered around Mexico City, prevailed. True cities evolved in the valley, flourishing on an agricultural system that provided 85% of the food, a figure as high as that of some contemporary rural societies. Social organization accelerated in complexity. Religion grew in influence. A standing army was created. Before the conquest of Mexico by the Spaniards in 1519, the Mixtec influence was superseded by that of the Aztecs, a socially advanced but more warlike people from the north.

We can draw some parallels between the New World and the Old. As in the Middle East, the evolution from a hunting and gathering way of life to one dependent upon agriculture as the energy base provided a more predictable and abundant food supply. Agriculture initiated a sedentary existence which led to population growth, the refinement of both the material and nonmaterial culture, and the subsequent rapid expansion of a relatively sophisticated civilization.

The Chinampas It is often assumed that these early systems of agriculture were primitive in terms of technological and ecological understanding. Nevertheless, the development of agriculture along the Rio Grande in New Mexico, and the agricultural systems in the vicinity of the area currently occupied by Mexico City, prior to the European invasion of the New World, illustrate that this assumption is not necessarily true.

The agriculture of the latter region was probably one of the most productive and ecologically conservative that has ever existed anywhere in the world, surpassing in yield—relative to the land area consumed—even our contemporary mechanized systems. Based on the *chinampa,* a long narrow strip of land surrounded by water on three sides, this farming technique has persisted to some extent until the present in the area of Xochimilco, south of modern Mexico City.

A properly maintained chinampa will yield several crops per year, remaining fertile without having to lie fallow for a part of the year. Soil on the strips is replenished by dredging the canals, to obtain the rich muck deposited in them. The chinampa system originated when the Mixtecs were draining Lake Texcoco, and was adopted by the Aztecs when they assumed power through conquest. The high productivity of the chinampas facilitated the Aztecs' dominance over the whole region, a feat accomplished in a relatively short period.

Chinampas can be enormously productive. When the Spanish conquerors arrived at Mexico City in 1519, they found a region of agricultural plenty. The Aztec emperor was receiving annual tributes of food, in the amount of 7000 tons of corn, 4000 tons of beans, 5000 tons of chilies, 3000 tons of cocoa, two million cotton cloaks, a ton of gold, and two tons of amber and feathers of the quetzal, a bird presently threatened with extinction because of habitat destruction and a thriving demand for its long, iridescent green tail feathers. These figures are particularly impressive when one considers that two centuries earlier the Aztecs had been semi-barbarian and materially poor. When they originally migrated into the Mexico City valley, they lost battles with their neighbors and retreated to two small islands in Lake Texcoco. Here they copied the chinampa system already practiced by the Mixtecs. In a relatively brief period, large food surpluses were garnered, enabling the Aztecs to overcome rival groups.

In understanding the development of this unique agricultural tactic, the chinampa, it is useful to know at least a few details of the geography of the Mexico City valley. A land-locked basin encircled by mountains, it lies more than a mile above sea level and has an area of some 3000 square miles. In pre-conquest times, a shallow sheet of water, formed from the drainage of the rivers and mountain streams, covered the floor of a fourth of the valley, forming Lake Texcoco. During the wet season, the large but shallow lake was known as the Lake of the Moon. During the dry season, it shrank into five smaller lakes. People had found the marshy, fertile valley by 2000 BC.

From the time of the conquest of Mexico by Europeans, the valley has been changing continually. In the colonial era, much of the land was drained off for European-style agriculture. More water was lost when Porfirio Diaz cut a great tunnel out of the valley to the north in 1900. Further drainage has occurred from the tapping of underground water supplies for the evergrowing population of Mexico City, which today is one of the most densely populated cities in the world. The population is nearly 15 million. Intensive deforestation of the surrounding mountains has caused erosion and resulted in accompanying water loss. Today, the valley is somewhat dry and dusty, teeming with people, and bearing little evidence of the agricultural miracle which once was there. The remains of the chinampa system are intact at Xochimilco, maintained largely for tourism and for the cultivation of flowers and some produce for Mexico City's markets. The chinampa, the so-called "floating gardens," are a small remnant of an agricultural method once among the most productive in the world.

The particular physical qualities of the valley were responsible for the inception of chinampas, a system that could not be reproduced in many locales. The high altitude may have been an additionally beneficial factor, reducing the number of insects which plague farmers in the humid jungles of the lowlands.

Mayan Farming Techniques   A traditional assumption has been that the tropical lowland agriculture of the pre-Columbian *Mayas* was of the *swidden* variety, a shifting cultivation in which farmers would deforest an area, grow crops in it for a few years or less, then move on to another area, allowing the former land to lie fallow. This practice, also known as "slash-and-burn," is still very common in rural areas of the American tropics. Recent interpretations of the Mayas' farming techniques, however, question this theory (Turner, 1974).

Ancient forms of intensive cultivation have been discovered in parts of the American tropical lowlands where it was previously thought only limited swidden agriculture was practiced (Denevan, 1970). Also, the theory of agriculture as a response to growing population pressures gives us another perspective (Boserup 1965). Studies suggest that certain regions within the boundaries of Maya occupation supported populations too dense for the carrying capacity of the primitive agricultural systems. The numbers of people may not have been large by existing standards, but there were too many to be sustained by swidden farming.

Research in the Rio Bec region of the southern Mexican states of Campeche and Quintana Roo, during 1973, supports the theory that the Mayas were practicing a more sophisticated agriculture than the swidden system. Terraces and raised fields were found throughout the area of Mayan habitation, including the Mayan Mountains of Belize, a locale formerly ignored by investigators, despite proximity to the major population center, Tikal.

The Rio Bec area, which has been the focus of recent studies, encompasses southeastern Campeche and proximal parts of Quintana Roo. Geographically, it offered a variety of local environments for Mayan cultivators. Dominated by limestone hills 20 to 60 meters high with steep slopes, the uplands contain well-drained but shallow soils. These vary from 5 cm (centimeters) to 45 cm in depth, and are highly subject to erosion during the torrential rains of the wet season. Level terrain varies between wooded areas and small savannas; drainage here is limited by a thick underbase of gray clays. So for permanent cultivation of the uplands or lowlands, two considerations are foremost: preventing erosion in the former, and draining the latter.

Since there are tens of thousands of relic terraces crisscrossing southern Campeche and Quintana Roo, in an area of about 10,000 square kilometers, there can be little doubt that terracing was a major agricultural tactic. Most of these terraces were built on the slopes of hills and were embanked with rocks. Rainfall provided the water; there was no form of irrigation. The annual precipitation in this region is high, so the impetus for developing complex irrigation systems would have been minimal. On each level, eroding soils from the slope above were collected behind the lower terrace wall, so topsoil was thinner directly below each embankment.

A less common design was the check-dam terrace found in ravines. These resemble the channel bottom weir terraces in various parts of the Mayan lowlands. In both, the purpose is to capture the fertile silt present in runoff, by laying the terraces across runoff channels.

Terraces to the east of Becan usually consisted of cut limestone, backed by a fill of rubble. To the west, the embankments tend to be smaller and lack a defined area of rubble fill. They are of dry-laid (without mortar) limestone, or of upright rock slabs sunk 20 cm or so into the ground. This latter terracing bears a notable resemblance, especially in its decaying state, to "linear border" relic terraces in the southwestern United States.

In all probability, these terraces were constructed to repair the loss of shallow, invaluable layers of topsoil through erosion. Continuous deforestation of the slopes, with frequent exposure to torrential rains, caused serious erosion problems during the Mayan occupation, as it still does today. It is interesting that current agricultural practices in areas of Mexico and in much of Latin America are far less conservative of indispensable topsoil than were the Maya's terrace designs. Today, corn is frequently grown on steep slopes, and no terracing of any kind is done. This lack of forethought promotes severe erosion which can make the land worthless for agriculture within only a few years.

In the Mayan system, topsoil captured by the terracing embankments formed layers from 25 to 45 cm thick. A topsoil layer may have served one or all of the following purposes: a more fertile medium for a dominant crop, such as maize; a layer of adequate depth for the cultivation of root crops; or it may have been spread evenly over the entire terrace, instead of being left in graded depths. In whatever manner it was used, the topsoil was at least retained for farming instead of being washed away down the hillside.

During Maya times raised fields, also known as ridged or drained fields, were commonly employed in the lowlands in seasonally flooded areas where the soil was rich but poorly drained. Since the tropical rain season may last for as long as four or five months, the Mayas constructed raised dirt platforms interconnected by drainage canals. (Note the similarity here to chinampas.) Advantages of such a system include the easier control of weeds, a greater concentration of organic matter and topsoil, and easier harvesting.

The existence of both terraces and raised fields testified to the Mayas' intensive agriculture. These people practiced continuous farming rather than allowing the land to fallow, an indication of population pressures increasing during the Late Classic period (627–757 AD). That there was such continuous agriculture makes one doubt the swidden-collapse theory explaining the demise of Mayan civilization, at least as applied to the Rio Bec region. If terraces and raised fields are

Llamas are one of the few New World animal domesticates.
Llamas are used in the highlands of South America for wool, as
beasts of burden, and occasionally for meat. (FAO photo)

found to be equally common throughout other areas once dominated by Mayas, the theory may have no validity in terms of the total Classic Maya civilization.

Other explanations of the collapse of Mayan civilization now appear more likely. Excessive deforestation may have culminated in short-term but broad environmental problems with which the Mayas could not cope. Or they may have been faced with "monoculture disasters," an even greater probability. In the latter case, dependence upon one or a few dominant cultivated plants, as well as rigid farming techniques, creates a situation of minimal flexibility if specific diseases, pests, or climate changes destroy the major crops. The entire food base of the Mayan empires may have been susceptible to the ravages of a crop-specific disease able to destroy within a few years the foundation of their civilization. Once a secure food base had been obliterated, there was no way in which the sources of ruling authority could be supported; the empire dissolved.

The Environmental Impact of Agriculture   We cannot be certain of the extent to which human beings' agricultural practices altered the environment during prehistoric times. Our conjectures must be limited to inferences. Some effects were undoubtedly fleeting; whereas others were permanent. We do know that modern agricultural techniques, aided by extensive mechanization, the use of pesticides, and fertilizers, can cause profound environmental damage. However, it would be a mistake to presume that more primitive methods did not make lasting marks on the environment.

An exemplary case is the slash-and-burn agriculture of the aboriginal Indians of South America. Laced through the jungles of northern South America are savannas, grasslands harboring small pockets of trees. Some investigators believe that these grassy expanses arose in response to swidden agriculture over the course of millennia. The continuous cutting out of virgin rain forest gave the grasses an improved foothold. Once established, the grasses formed a thick mat of sod preventing most trees' seeds from taking hold, except in scattered niches. Thus the forest was prevented from regaining the savannas.

Land in cultivation has, necessarily, been radically altered in terms of the ecosystems it once supported. The monocultural nature of croplands excludes the proliferation of other species. Accordingly, wildlife diversity is reduced. This is also true of commercial forests, or tree farms, which in no way resemble a natural, balanced ecosystem. Domesticated animals may also cause striking environmental changes, by competing with native wildlife for food and by denuding the land of its vegetation. In the American West, great areas once supporting lush grasslands are nearly deserts now as a result of overgrazing by sheep. The introduction of domesticated animals—especially goats—into the Galapagos Islands, off the coast of Ecuador, caused havoc among the isolated indigenous fauna, bringing about the extinction of several species that were not at all endangered prior to the arrival of human beings and their livestock.

Extensive environmental changes resulting from agriculture may be essentially transitory. However, it is doubtful that any ecosystem, once destroyed, will regenerate exactly as it was before. Towering jungle engulfed the Mayan center Tikal after it was abandoned—replacing entirely the fields and settlements carved from the forest during Mayan rule. In fact, the Peten area surrounding Tikal, where extensive farming was once practiced, is today one of the last remaining wildernesses of any size in Central America. And before excavation, Angkor Wat in Cambodia was hidden by a cover of jungle.

Nevertheless, it is probably true that human agriculture has, during the thousands of years since its inception, produced greater changes than one might suppose. Certainly all environments naturally undergo changes. However, the influence of human beings may have accelerated those changes and altered courses. Parts of what is now the Sahara Desert were reduced from productive, forested regions to wasteland through deforestation and abusive agricultural practices. Perhaps the most significant element of agriculture is reducing tree cover which can lead to changes in the rainfall pattern of a region—lessening the quantity of rainfall. Once this happens, the land may evolve—at times with drastic rapidity—into a desert.

There is no doubt that human beings have made a more lasting mark on the world than any species before them, and that our influence today is by far the most reaching. Mechanized agriculture, while it does facilitate the production of great quantities of food, may have the unfortunate side effect of causing the loss of large quantities of topsoil. Fertilizers and insecticides, when used extensively, tend to have adverse environmental effects as well.

Existing problems are not irresolvable, but the dilemma with which we are confronted is that of a burgeoning human population which will require an ever-increasing quantity of food. In attempts to meet the need, more land will be taken over for agriculture, and one cannot help but wonder at the outcome if the world population is not stabilized (see Unit 5). From a biological perspective, environmental diversity is beneficial in any ecosystem. The consequences of over-cultivating the world may be regrettable. Our goal must be to achieve an ecological balance while meeting the food requirements of the world's peoples.

Suggested Readings

Braidwood, Robert J. "The Agricultural Revolution." *Scientific American* 203 (1960): 130–148.

Denevan, W. M. "Aboriginal Drain-Field Cultivation In The Americas." *Science* 169 (1970): 647.

Iverson, Johannes. "Forest Clearance In The Stone Age." *Scientific American* 194 (1956): 36–41.

Mangelsdorf, Paul C.; MacNeish, R. S.; and Galinat, W. C. "Domestication of Corn." *Science* 143 (1964): 538–545.

Solheim, Wilhelm G., II. "An Earlier Agricultural Revolution." *Scientific American* 226 (1972): 34–41.

Turner, B. L., II. "Prehistoric Intensive Agriculture In The Mayan Lowlands." *Science* 185 (1974): 118–124.

In Unit Three, plants of major cultural and economic significance to human beings are considered. First, we examine the origins and impact of the "seeds of civilization": wheat, corn, rice, and several other grains. We then detail the harvesting and processing of various fibers and tobacco. Important fruits, roots, vegetables, legumes, and nuts are next discussed. In chapters 10 and 11 we focus on the relatively recent development of sugar and latex production and the long-established fascination with spices and essential oils. Lastly, we find that plants produce many of our most popular beverages: coffee, tea, and cocoa, for example.

UNIT 3

CHAPTER

6

THE SEEDS OF CIVILIZATION: WHEAT, CORN, RICE, AND OTHER GRAINS

Wheat   Wheat, as discussed in the preceding chapter, is a basic food crop throughout the world. A species of grass, domesticated wheat is a major cereal crop, second only to rice; together, these two constitute the basis of 40% of all human energy. In 1970, 288 million metric tons of wheat were grown worldwide (a metric ton is 2204.6 pounds). The USSR leads in world production, followed by the United States and Canada, Central Europe, Turkey, Argentina, China, Australia, and India. Today, the American wheat belt is as important as it was in the 1930's, yielding more than 30 million tons of grain annually.

The oldest known remains of wheat were discovered at Jarmo in Iraq, dating to 7000 BC. By 5000 BC, wheat was being cultivated in the Nile Valley, and had been established in the Euphrates and Indus valleys no later than 6000 BC. There is evidence that it was present in China by at least 2500 BC and perhaps earlier. It was already being grown in the British Isles at approximately the same time or somewhat later.

Wheat is well adapted to harsh climates. It is cultivated on windswept prairies too dry for rice or corn. The crop requires moderate moisture and cool weather for early development, followed by a dry summer for maturation. Tropical areas are not conducive to growing wheat. The major disease, the wheat rust fungus, (Puccinia graminis) thrives in warm, humid climates. Thus wheat has never become established significantly as a crop in moist regions.

"Wheat" refers to a complex of several species, the most common of which is Triticum aestivum. Two of these species are diploid (having the basic chromosome number doubled), nine are tetraploid (having twice the diploid number), and four are hexaploid (six times the basic chromosome number). One species, Einkorn wheat (Triticum monococcum) is unusual in that it produces only a single grain in each spikelet. Although yielding a sparse harvest, it will grow in poorer soil which would not support many other crops. The remains of T. monococcum have been uncovered both at Jarmo and in Swiss lake dwellings.

Modern bread wheat (Triticum aestivum) exists in many varieties due to its widespread cultivation, but it can be separated broadly into two groups related to the mode of cultivation: winter wheats and spring wheats. Spring wheat is usually sown in the early spring and harvested about four months later. These varieties are grown in colder regions of the United States, in Russia, and in Canada. Where there are abundant winter rains and temperatures are less severe, winter wheats may be grown. These are planted in the fall to germinate and grow some before winter sets in; and are harvested the following midsummer.

During the initial stages of agriculture, an average wheat yield was probably 4 to 6 bushels per acre. Presently, a typical yield is about 25 bushels per acre, although good management and the use of hybrid varieties can increase production to as much as 100. Development of the Gaines hybrid from dwarf Japanese varieties resulted in a yield of up to 100 bushels per acre, particularly in the Pacific Northwest of the U.S. This hybrid was essentially responsible for the "green revolution." Called the Norin-10 germ plasm, it sparked research leading to the development of certain varieties in Mexico, for which N. E. Borlaug received the Nobel Peace Prize in 1970.

From the Wheat Flower to Bread   The flowers of these exceptional varieties are usually perfect, numbering from one to five in each inflorescence (the spikelet) and are self-pollinating. The mature ovary, or fruit, is a composite of a small embryo (the germ) with a single cotyledon (seed leaf) embedded in the starchy endosperm. This is surrounded by protein-storage cells, called the *aleurone*, all of which is encased in the *pericarp*. Upon processing, the aleurone and pericarp became the bran of a grain.

In an average grain of wheat, the endosperm comprises from 82% to 86%, the pericarp 5%, the aleurone from 3% to 4% and the germ 6%. Analyses of the nutritional content of wheat will differ from one strain to another, but average figures are: 14% protein, 2% to 3% fat, and 78% carbohydrates, with the remainder being minerals and moisture.

Wheat flour is suitable for making bread because of two binding proteins, *glutenin* and *gliaden*. The higher the content of these proteins, the stickier the dough will be. Without wheat or rye flour, it would not be possible to make most of the breads and the great variety of pastries and other related foods so abundant in Western cultures. Flour from other grains, such as corn, would not serve in the preparation of these foods.

Milling of wheat involves cleaning the grains, moistening them to toughen the bran and prevent it from fragmenting, crushing the grain, separating the bran and germ from the endosperm (starch), and further grinding and bleaching of the flour. Some protein and vitamins are removed in the germ and bran, which are fed to livestock in many parts of the world. White bread contains smaller quantities of protein. The wheat germ, sold as a cereal, is more than 30% protein. There

Above Bread, although decreasing in importance in the daily diets of people in developed countries, is still the "staff of life" for most inhabitants of the developing countries. (FAO photo)

Left An Andean Indian farmer in his field shows the good results achieved by using selected wheat seed and fertilizer. (FAO photo)

is one practical reason for eliminating the germ during the milling. The flour will keep longer with a reduced protein and fat content. Modern roller milling also affects other constituents. Roller milled wheat retains 6% to 16% of its thiamin, whereas more primitively stone ground wheats retain 60%. Minerals, vitamins, and soy flour are added to some breads during commercial production to compensate for these losses.

In making most breads, another plant—albeit a microscopic one—is an integral component: yeast, an *ascomycete* (of the class *Ascomycetes*), a unicellular fungus. Added to a glutinous dough consisting of flour, water, oil, and sugar, the yeast cells multiply and, in the metabolism, release carbon dioxide gas. The gas forms bubbles that cause the dough to "rise." Baking coagulates the glutenin protein, and kills the yeast cells. This results in setting of the bread. Unleavened bread is bread made without yeast, so it does not rise.

Experimentation with wheat varieties is ongoing. A hybrid of wheat and rye, known as *Triticale,* has been developed. Although the initial cross is sterile, it can be made fertile by *polyploidy,* (multiplying the basic chromosome number). This variety is used in the USSR. Its protein content is higher than that of wheat, and it is more winter-hardy—an important consideration in Russia. However, at this time the yields are less than with wheat. Further genetic engineering may transform it into an economically important crop.

Corn

Unlike domesticated wheat, the progenitor of which is generally agreed to be a wild wheat, the origin of corn is uncertain. Although Mangelsdorf has offered convincing evidence that corn is descended from a species of wild corn (see Chapter 5), several investigators disagree with his conclusions and believe that corn (*Zea mays*) is the descendant of teosinte (*Zea mexicana*), the closest wild relative of maize.

Teosinte (*Zea mexicana*) is an annual grass which looks much like maize (corn). In the wild, maize and teosinte hybridize freely when isolation barriers do not exist. In fact, this is what happens in the Sierra Madre Occidental of northern Mexico, the Central Plateau and Valley of Mexico in central Mexico, and in Huehuetenango in northern Guatemala. *Tripsacum* does not hybridize readily in the field with maize, but will do so in experimental circumstances. Corn is dependent for its heterotic vigor on the wild populations of teosinte which grows sometimes in and near maize fields. The extinction of these native teosinte populations would be disastrous in terms of the well being of the corn stock of the future. With extinction of the teosinte, the necessary alien germ plasm would not be introduced into native races of maize. This would be disastrous because maize species depend on hybridization to maintain yields. However, the extinction of teosinte seems to be taking place (Wilkes, 1972).

Human beings and maize have had an extended relationship in which there has been conscious and unconscious alteration of corn's original genetic structure. Today, maize can no longer reproduce itself without being cultivated. Some cultigens, such as wheat, can reproduce in the wild. However, because corn is not represented in a wild state, there has long been speculation about its origin (see Chapter 5). Even pre-Columbian Indians knew that the plant did not grow wild in the countryside. The Nauhautl believed it had been proffered by red ants.

> Once again the gods asked, "Oh you gods, what is man to eat?" And a search was begun high and low for a food. It was then the red ant brought back corn seed from the land of plenty (the underworld). (Garibay, K. 1945; in Wilkes, 1972).

Today, teosinte grows naturally in the seasonally-dry subtropical zone having summer rains. This conforms to the climate of the cultural area of Mexico and Guatemala referred to as Mesoamerica. The appearance of teosinte is so close to that of corn that distinguishing two plants growing side by side in a field may be difficult. The surest way to distinguish between the two is by the female infloresence, which has a distichous* spike in teosinte and a polystichous† structure, the "ear," in corn. In teosinte, the grains are dispersed as individual rachis segments from the disarticulated spike. Maize cannot disperse its own kernels, but is dependent on human beings for propagation.

Hybridization between maize and teosinte is common. As a result, the teosinte is becoming more and more like corn. It is cleared away from ditches surrounding the corn fields; but in the cornfields it so closely resembles the crop plants that it is indistinguishable from them until the time of flowering. Usually the teosinte is not removed until pollen has been shed. By that time it has been able to cross with the maize, and mature grains have been formed in the teosinte, on the first flowering pistillate spikelet. Although farmers make concerted efforts to eliminate teosinte from their fields, they use the plants as mulch or feed them to their livestock which pass the resistant,

*divided into two segments or vertical rows
†arranged in more than two rows

indurated seed out in their feces. The manure is later used as fertilizer for the corn field. In this sequence, it would seem difficult to remove teosinte as a companion plant of corn.

However, the extinction of teosinte may be realized; and if it is, the evolutionary pattern of corn hybridizing with teosinte will cease. Corn and teosinte seeds can be stored in seed banks, but the genetic intermingling of teosinte and maize would cease. The formation of new races in maize as a result of that intermingling cannot be preserved. How would the destruction of teosinte plants outside of the fields come about? One way would be through the use of weed killers, and the replacement of native maize crops with commercial hybrid seeds. According to Wilkes, the geographical distribution of teosinte is today about half of what it was in 1900. With the need for abundant food crops so great, the loss of the genetic variability of corn would be unfortunate. In light of the role teosinte has in hybridization of maize, it would be astute to preserve the plant's genetic material for future studies with modern corns. Many Mexicans have long perceived an intimate relationship between these two plants, believing that after a few generations, teosinte turns into corn. Of course it does not, but one can understand this misconception, so marked is the similarity.

Corn is monoecious, with male and female flowers on different parts of the plant. Under natural conditions it is pollinated by wind, generally resulting in cross-pollination. However, a first step in the production of hybrid corn is the isolation of inbred strains. This is accomplished by artificial self-pollination. In plants such as corn, in which the male and female flowers occur on the same part of the plant but ordinarily do not pollinate each other, such artificial self-pollination is genetically three times as effective in producing homozygotes as brother-sister mating in animals.

Agronomists create these extremely inbred strains in order to breed new forms, to isolate genetic characteristics considered desirable and to eliminate those that are impractical or deleterious in terms of crop production. Many genetic deficiencies (such as the loss of chlorophyll, weak stems, and dwarfed strains) become apparent through inbreeding, which causes characteristics which previously were not apparent to manifest themselves. Because such characters are recessive, these less desirable qualities in the genetic makeup may not be evident until brought out by inbreeding. The breeder can then remove these qualities from his or her own plants. Inbred selection results in a very uniform corn crop after only five or six generations. However, uniform strains yield approxi-

*Hybrid corn improvement programs are being
started in many countries. In Cameroon these
students are practicing selecting the best ears of
the crop for hybrid seed production. (FAO photo)*

pollinated varieties. The function of such strains is that of potential parents of hybrids exhibiting superior features and also high yields.

Characteristics sought in corn breeding experiments include: stiff stalks, which will support more ears without breaking; a greater number of ears per plant; drought, insect, and disease resistance; any other qualities needed for more nutritious, hardy, and productive plants. Hybridization is responsible for a marked increase in the yields of many food plants and is the key technique behind the "green revolution." For example, in 1930 an average corn crop was about 22 bushels per acre; today a harvest of 230 bushels per acre is not uncommon.

The hybridization of corn or of any other crop is not without liabilities. There is an accompanying loss of genetic diversity, leaving the crop plant more vulnerable to the attacks of new pathogens. This is what happened in the southern corn blight of 1970: In seed corn production, it used to be necessary to detassle the corn to prevent open pollination. To solve this problem, strains with a male *cytoplasmic sterility* factor were developed, so male and female parent plants could be grown in the same field. Then, in 1970, a blight wiped out vast quantities of corn crops in the Midwestern U.S.—selectively attacking the plants with a cytoplasmic sterility factor. Genetic diversity, which may be limited in hybrids, is insurance against this kind of agricultural disaster.

Through an overdependence on hybridization, there is the potential for losing much of the genetic material from natural corn. In the United States particularly farmers are reluctant to cultivate open-pollinated varieties. As a result these varieties are grown almost solely at USDA laboratories. The grains are stored in anticipation of future need. It is also important to have reserves of corn indigenous to Latin America and other countries. Because hybrid corn has little genetic plasticity, it is conceivable that genetic material from open-pollinated varieties will be needed at some future time to improve, or even to save hybrid strains. At best, the potentials of hybridization are not limitless. Some botanists have been making concentrated efforts to discover wild relatives of cultivated crops—hoping to find new genomes to introduce into our genetically specialized cultigens.

The success of hybrid crops is greatest in the first generation. Second-generation progeny are much less productive, and uniformity diminishes in succeeding generations. Thus, farmers must buy new hybrid grain every season. Growing hybrid grain is an extensive business. Currently, there are

Rice involves much hand labor
from cultivation to harvest.
These scenes are from Sri
Lanka.

Above *Transplanting rice.*
(FAO photo)

Left *Sowing rice. (FAO photo)*

hundreds of hybrids which are genetically adapted for a variety of environmental conditions.

Corn has been historically and remains today the basic food plant of the Americas. Comparable to wheat in the Middle East, a highland origin of about 4500 feet has been postulated for corn. The plant was adaptable to a variety of situations. By the time of the European invasion of the New World, corn was being grown from southern Canada to southern South America. It did not achieve much popularity in Europe, probably because the Europeans did not know how to prepare it palatably where it produced a crop.

As a primitive food, corn was probably popped, parched, boiled, and ground. The Indians also soaked the grains in wood ashes to make hominy. Today, corn appears as a food product in Latin America primarily in the form of *tortillas*, which are flat, round cakes made from *masa* (cornmeal). Tortillas are served with almost every meal and may be thought of as the "bread" of Central America.

Rice

Of the three principal grains on which most civilizations have been founded, rice may have been the first to be cultivated. Today, rice is a dietary staple for more than 60% of the world's people. About 80% of the world rice crop is grown in Asia and India. Rice is also important as a food plant in Africa and tropical America; perhaps, in part, because of the influence of East Indian populations in those areas. Cultural variation in the use of rice is reflected in the American consumption, an average of 8 pounds per year per person, compared with the Asian consumption of up to 400 pounds per year per person.

Nutritionally, rice is somewhat inferior to corn and wheat, having a protein content of only 7.5%. Improved hybrid strains of rice usually offer more protein, but the yield tends to be lower with these varieties. The nutritional value may be improved through fermentation, as the presence of yeast will increase the amount of protein. Fermented rice is commonly eaten in India and areas of South America where there are large East Indian populations, but it is not popular in the Orient. It tastes and smells like slightly sour milk, is easier to cook than unfermented rice and is nutritionally improved.

As with wheat, the processing of rice removes much of its nutritional content. After milling, polishing, and drying, all of the vitamins C and B have been destroyed. (However, vitamin C is not retained in any dried grain.) "Polished" rice is composed of approximately 92% carbohydrates; only 2% of the additional materials are of other nutritional value. In comparison, whole rice still has the aleurone; and, consequently, vitamin B and a greater percentage of protein.

Paddy Rice   In Asia, rice is usually cultivated by arduous, time-consuming manual labor. The culture of Asian rice is known as *paddy*, a method applicable to only one type of rice production. Paddy rice requires flooding, as it achieves much of its growth in standing water. Another type (highland rice), from which paddy rice is probably descended, grows without flooding. Upland strains are less productive than paddy rice, but require less effort in cultivation.

Paddy rice is customarily sprouted in wet seed beds and then transplanted to flooded fields after the onset of the wet season's heavy rains, which usually occur between November and January in southeast Asia. The paddies are weeded by hand, to protect the young plants from the vegetative competition that exists in tropical areas. After two months, the standing water is drained. When the plants are mature, the grain stalk is cut by hand and allowed to dry. Threshing is also done manually, usually by women. In some areas the rice is treaded upon, either by cattle or by humans.

Paddy culture, although time-consuming and laborious, can be ecologically beneficial. Nitrogen is supplied by the blooms of blue-green algae, which fix nitrogen into the muck of the paddy soil. Carp and other fish frequently invade or are introduced into the flooded areas. The fish grow to maturity in the paddies and add their nitrogenous wastes to the benefit of the rice crop's production.

The introduction of chemical fertilizers has upset the balance of the paddy culture ecosystem. Overfertilization results in the growth of great quantities of algae; algae death and decay consumes oxygen. The lack of oxygen may prohibit the raising of fish, an important potential protein supplement for the rice farmers.

The Modern Cultivation of Rice   In Japan, the cultivation of rice has undergone extensive mechanization. In most of the rest of Asia, the softened ground of flooded paddy fields is still tilled by primitive means, such as oxen pulling forked sticks. In present-day Japan this is often accomplished with small, mechanized hand tractors. Yields of approximately 4000 pounds per acre, or four tons per hectare (a metric unit of area equal to 2.471 acres) are common. Chemical fertilization is a standard part of the cultivation process. Many Japanese farmers also use a green manure, a legume grown in the paddies after drainage and plowed into the soil before subsequent plantings. Often, chemical fertilizers are added as a supplement. After the rice is planted, still more fertilizers are intro-duced. Upon maturity, the rice is harvested and threshed by portable machines carried to the fields by hand.

Since Japan's agricultural land is severely restricted in area, cultivation is intensive. Recently, the rice paddies have been irrigated extensively, so farmers can use the land to grow other crops during the dry season. This technique has had two results: The total crop that can be cultivated per unit of land is expanded. Also, it has been found that crop produc-tion during the dry season tends to be superior to the yield during the rainy season—perhaps because of warmer temper-atures during the dry season.

In the United States, rice cultivation is highly mechanized. Fields are irrigated mechanically, and when the rice is mature it is harvested by combine. Rice was introduced in the Car-olinas in 1647, and quickly became a significant plantation crop due to the availability of slave labor for crop cultivation. Following the abolition of slavery, most rice was imported into the United States until mechanized agriculture solved the labor problem. Currently, the major areas of rice cul-tivation in the U.S. are California, Texas, South Carolina, Arkansas, and Louisiana.

Much effort has been made to develop new hybrid rice strains. The International Rice Research Institute (IRRI) in the Philippines was established in 1962, with the cooperation of Australia, the United States, the Philippines, Japan, India, Taiwan, and Ceylon. The Institute's purpose is to enlarge the potential of rice crops, in terms of productivity and other desirable qualities, such as disease and insect resistance. Investigators attempt to breed strains having immunity to rice blast fungus, an ascomycete fungus which is one of the crop's principal diseases. Such cooperative agricultural research ventures are urgently needed in view of increasing global food shortages. Many of the countries dependent on rice for the sustenance of their people are small. These nations lack the necessary capital to invest in research necessary to breed new strains of crops. Cooperative efforts are their main hope.

Various Uses of Rice   In Japan rice is used in several ways other than direct consumption as a grain. Rice flour is utilized as a food starch, and was once used as a cosmetic, although the latter function is today limited primarily to *Kabuki* actors. It is also tradi-tionally the face powder of the bride during a Shinto wedding ceremony. Matting is made from the rice plant straw; and in the countryside the straw was used in the fabrication of rain-coats until recently when plastic products became available.

Often considered an Asian crop, rice is grown increasingly in many parts of South America and Africa. (FAO photo)

A fine field of barley. (FAO photo)

A Japanese wine called *sake* is made from fermented rice. The importance of rice to the Japanese is evidenced by the fact that, even in modern industrialized Japan, corporate executives will donate the first sake and samples of the first rice crops to Shinto temples. The gesture is considered to be good luck. The lee from sake are used to pickle vegetables and fish, being rather sweet and having considerable nutritional value.

Rice is one of five plants held sacred by the Chinese and Japanese. Each year it is planted by the Japanese emperor, as an acknowledgment that rice is a gift from the Shinto gods. The Chinese word for rice means "grain of good life." The Japanese term (*gohan*) is a contraction of two words: *go* (five) and *hana* (flower), indicating that primitive varieties probably had few grains per head.

Common myths about rice include the belief that "rice paper" is a rice product. Actually, it is made from another plant (*Tetrapanax*). Another belief is that a continuous diet of brown rice will lead to spiritual liberation. In fact, use over an extended period would more probably cause severe nutritional problems related to insufficiency of protein and other nutrients.

Other Grains   Barley.   This grain is a complex of several grasses of the genus *Hordeum* and was a companion plant to wheat during the rise of civilization in the Near and Middle East. A hardy group of grasses, barley is adaptable to high altitudes, resistant to cold, and will thrive even when the growing season is shortened through adverse conditions such as drought. However, as with wheat, they cannot tolerate heat and humidity.

One of the most ancient of cereal crops, barley was probably first cultivated by people in the Near East, where two species of wild barleys still thrive: *H. spontaneum* and *H. maritimum*. Abyssinia and Tibet have also been suggested as focal points of the origin of cultivated barley. As early as 2800 BC, barley was used as a food crop in China; for millennia it had been consumed by people and beasts of burden in Egypt.

Today, barley is the fourth most important crop economically, after those discussed previously. A total yield in excess of 100 million tons is produced annually, with the Soviet Union growing the most, followed by France, the United Kingdom, the United States, Turkey, India, Morocco, and Argentina. The contemporary use of barley is primarily as animal food. About a third is used in the making of malt, a component in the production of beer and other liquors, and also an element in confections and a variety of the health

Above *High yield sorghum. (FAO photo)*

Left *Harvested millet. (FAO photo)*

foods currently so popular in the United States and Europe.

Natural barley has a reasonable protein content (13%); and there is evidence that, by increasing the frequency of mutation, new strains may result having a considerably higher protein content (Doll et al. 1974). Experiments yielded strains having 40% more lysine in the protein than the parent plants had. However, there was a related reduction in crop yield of 10% to 30%. But the most promising mutant had the least loss of yield. Further experimentation may solve this problem, as well as the problem of the mutants' having a slightly reduced digestibility.

Oats. A species of grasses of the genus *Avena*, oats are grown today primarily as livestock feed, although the grain does comprise popular cereals for human consumption as well. They are widely grown in Europe and the U.S., being well suited for diverse soil types, climatic variation, and differing cultivation techniques. Currently, world production of oats is relatively not high, totalling about 50 million tons annually.

The original domestication of oats is obscure, probably occurring at different time periods in the areas now known as the Soviet Union, Africa, and the Near East. The cultivation of oats is fairly recent. Classical Greek writers acknowledged the plant's existence, but as a weed, not as a crop plant. The common species of today, *A. sativa*, is not found in the wild. Some investigators believe it to be a mutation from *A. fatua*, a wild oat species found in Asia Minor and southeastern Europe. The cultivation of oats probably began approximately at the beginning of the Christian era.

Rye   Rye (*Secale*), as well as oats, was considered a weed in the fields of wheat and barley in ancient times. Today, rye is principally a European-Soviet grain, with the latter heading world production. The total annual rye yield is about 35 million tons. The domesticated species, *S. cereale*, is not known to occur in the wild, and may have been generated from existing wild ryes, such as *S. montanum* or *S. anatolicum*, of the Mediterranean and Asia Minor.

The grain is reasonably high in protein content, approximately 13%. It is glutinous like wheat, and can be made into a bread flour. Rye is the staple of European dark breads. On the whole, this grain is not popular as a human food and is grown predominantly for livestock. In some areas of the world, straw from the rye plant is also used for thatching, various stuffing material, and paper making. The plant is sometimes employed as a cover crop and as "green manure." Rye may be fermented, for alcoholic drinks or for industrial alcohol.

Sorghums. This group of grasses (genus *Sorghum*) is of greatest economic importance in northern Africa and Asia. In these areas sorghums have been cultivated for at least 4000 years. During the past few years their popularity has grown particularly in the U.S., which now grows the most sorghums. India, Argentina, Ethiopia, the Sudan, China, and Manchuria also contribute heavily to the yearly world total of somewhat more than 30 million tons. The first domestication of sorghums probably took place in Africa or, perhaps, Asia. The grasses were being grown in Egypt over 2000 years ago, but sorghums, have been cultivated in the U.S. only since the 1800's.

Sorghums are arranged in four generalized categories, which are grounded in economic perspectives. Sorghos, which have sweet stalks, are used for forage and syrup; broomcorns are grown for their umbelliform inflorescence; grass sorghums serve as hay, pasturage, and silage; grain sorghums produce food for livestock and humans. The grain sorghums exist in a great many cultivated forms—species or varieties, depending on the classifier's point of view. They cross readily, and all have the same chromosome complement (2n = 20).

Perhaps the most important feature of sorghums in relation to their economic significance is that they are extremely resistant to drought, resuming growth after severe dry periods (Wilson and Whiteman 1965; Whiteman and Wilson 1965). One reason for this may be the plant's extensive root system, contrasted with a small leaf area. Suited to warm as well as xeric or dry conditions, sorghums characteristically grown in areas too dry for maize and too hot for barley or wheat. Farmers have found these grasses useful as a quick substitute when a traditional crop is threatened with failure because of disease or unexpected insect damage.

Millet. Millet is a catch-all term for several genera of food grasses grown predominantly in the Far East and in Africa. The total annual world production, only around 15 million tons, comes chiefly from India, China, Pakistan, and Africa just south of the Sahara. Small-seeded, annual grasses, millets have been termed the "poor man's cereal"; for they are eaten largely by poorer classes in the countries mentioned above, and are a forage crop as well.

Due in part to the small size of the grains, millets have not been able to compete with other cereal crops or to assume any degree of importance in feeding the world population. But these grasses have long been present as a cultigen; in fact, for a much greater time than many currently more important cereals. Remains of domesticated millet from the Tehuacan Valley in Mexico were dated at 5000 BC. In agriculturally advanced nations, millets are used in animal feeds, especially those for poultry and other birds, and as a soil cover and for pasturage. However, about 85% of the world production of millet is consumed by humans.

Millets are nonglutinous and therefore unsuited for making typical breads, although a flat, heavy bread is baked from millet in southeastern Europe. The protein content is less than that of wheat, but there is more fat. Nutritionally, the grain is more or less on a par with rice. Whatever the negative factors, millet can provide more sustenance given conditions of poor soil and limited rainfall than most other crops, thus providing an important staple in less fertile regions of Africa and Asia.

Other Grasses.   Some additional grasses are economically significant as food crops in limited areas or as specialty items. "Wild rice," Zizania aquatica, was a dietary element among North American Indians when Europeans first began to settle on that continent. A grass (but not closely related to true rice), Zizania was important in the diets of American Indians of the Great Lakes region prior to the arrival of Europeans. The primitive method of harvesting wild rice left numerous seeds to assure a crop for the following year. When harvesting by mechanical means was introduced, most of the seeds were collected by that method; and the species was soon threatened with virtual extinction until protective legislation was enacted. Wild rice traditionally is not cultivated, but is collected from natural populations. However, since the 1950's there has been some cultivation in artificial paddies. Yields from cultivated acreage are seven or eight times higher than those of naturally-occurring stands. Indians still use it for food, and also sell it as a cash crop which brings relatively high prices. A pound of wild rice usually sells for five dollars or more. Other minor grains include Job's tears (Coix lacryma-jobi), a grass having large, lustrous white grains that are used ornamentally and also parched, boiled, or milled into flour. Crabgrasses (Digitaria) produce a small seed known as "finger millet." In Ethiopia, teff (Eragrostis abyssinica) is an important millet used as a livestock grain and for a flour from which large flat cakes, known as injara, are made. These are a staple food of the Amjara. Straw from teff plants is also used as a component in the building of mud huts.

Suggested Readings

Doll, H.; Koie, B.; and Eggum, B. O. "Induced High Lysine Mutants in Barley." *Radiation Botany* 14 (1974): 73–80.

Garibay, K. A. *Epica Nahuatl* (Ediciones de la Universidad Nacional autonoma—Mexico, D. F.) 1945: 5.

Heiser, Charles B. *Seed to Civilization*. San Francisco: W. H. Freeman, 1973.

Whiteman, P. C., and Wilson, G. L. "Effects of Water Stress on the Reproductive Development of *Sorghum vulgare* Pers." University of Queensland Papers, Department of Botany, V. IV. St. Lucia: The University of Queensland Press, 1965.

Wilkes, H. Garrison. "Maize and Its Wild Relatives." *Science* 177 (1972): 1071–1077.

Wilson, G. L., and Whiteman, P. C. "The Influence of Shoot Removal on the Drought Survival of Sorghum." University of Queensland Papers, Department of Botany, V. IV. St. Lucia: The University of Queensland Press, 1965.

CHAPTER

7

FIBERS, FORESTRY PRODUCTS, AND TOBACCO

Wood, grasses, and fibers have long been incorporated into human culture—providing the materials for housing, paper, clothing, and a variety of other artifacts, from food utensils to religious fetishes. Woods have supplied a versatile and aesthetic medium for art and furniture; as well as being the basic construction element of homes and other buildings in many parts of the world. Grasses have been a ready source of the fabric for clothing and basketry. Plant fibers such as cotton and sisal are still in wide use, despite the abundance of synthetics that have reduced somewhat their importance.

Fibers

A plant fiber is a single, slender tapered cell with an unusually thick cell wall and simple pits. Single-celled fibers are the main constituent of hardwoods. Bundles of fibers are also commonly found in the phloem, pericycle, or cortex of the plant. Chemically, it is the cellulose and lignin content that make fibers useful to human beings, giving the fibers the qualities of toughness and durability. Cellulose, the main constituent of all wood and other natural fibers, is composed of linked glucose residues.

Most synthetic fibers, such as nylons, polyesters, and polypropylenes, are composed of synthetic polymers made from petroleum sources. In 1968, synthetics accounted for more than half the fiber production in the United States (about 3.5 billion pounds), and production has doubled in the past five years. However, because petroleum is becoming scarcer and more expensive, the use of synthetic fibers may well decline. Cotton production has been maintained through the incentive of government subsidy. Today, cotton is still in wide use, and many garments are a blend of synthetic and plant fibers. Oil shortages will probably have the effect of increasing the demand for cotton and other natural fibers.

Plant fibers are often grouped according to their origins in the plant part. For example, *surface fibers* are obtained from the surface of leaves, fruits, or seeds, as with cotton. *Soft*, or *bast* fibers develop in bark tissue, such as the fibers of hemp and flax. *Hard fibers* are strands of support cells found in the leaves of plants such as sisal. Fibers are also sometimes classified on the basis of their uses: filling fibers, braiding fibers, cordage fibers, textile fibers, and paper-making fibers.

The processing of fibers is accomplished by various means. Surface fibers are customarily separated by *ginning*, the removal of the surface hairs from the seeds. *Retting* is the removal of fibers from woody tissue by soaking the plant (flax, hemp, timber, for example) in water to break down other tissue. The process of *carding* aligns the fibers, and *spinning* twists them together, to form rope, string or thread. *Weaving* is the final process, by which fibers are manufactured into fabrics.

Weaving is an innovation as basic to human culture as the wheel. Human beings have practiced the art for at least 10,000 years. Woven flax fibers from Swiss lake dwellings have been dated at that age, and flax and hemp were woven in Egypt at the same time. Palm leaf fibers are present in Tehuacan excavations dated as early as 10,000 BC; they were probably woven into baskets and other utilitarian items.

Cotton

The genus of cotton, Gossypium, is represented world-wide by 30 species. Most of these are diploid (2N=26), but at least two are tetraploid. Short-lived perennials, the plants have an average productive life of five years. Cotton supplies by far the largest harvest of any natural fiber. The total annual world yield is some 12 million tons. The United States grows the most, followed by India, Brazil, Mexico, Pakistan, and the United Arab Republic.

The antecedents of Western culture, the Greek and Roman civilizations, used little cotton—employing silk, flax, and wool for their fabrics. Cotton has long been economically and socially relevant in Arabic cultures, however. As industrialization increased, it began to replace wool in importance in European fabric production. One interesting side effect was a reduction of plague-harboring fleas which thrived in woolen garments. Cotton fabrics were not just a result of the industrial revolution, but were, in part, the impetus for it in 16th century England.

Cotton is still associated with the institution of slavery. Most of the slave labor in the United States prior to the War Between the States was devoted to growing cotton and sugar cane. As early as 1810, the abolitionist John Woolman refused to wear cotton clothing or to use any other cotton products because of the manner of cultivation. Although it is traditionally linked to the Southeastern United States, today cotton is grown predominantly in Texas, Arizona, and California. The yields in these areas are twice as great as those of the South, largely because of climatic differences. The absence of many insect pests because of dry climate, and adequate irrigation have contributed to the marked increase in cotton productivity in California. In the South, the method of cotton cultivation ruined vast acreages of farmland through overcultivation and the resulting exhaustion of the soil's fertility.

Today, the planting of cotton usually begins during the first week in March. About 50 pounds of seed per acre is required. The young plants are thinned repeatedly and fortified with up to 600 pounds per acre of fertilizer. Also, an average of 50

Above *Afghanistan exports cotton to India, Japan, and Russia. The Food and Agriculture Organization (FAO) of the United Nations has sent experts to help the country improve the quality of its cotton, which, due to variation in staple length, had become almost unmarketable. (FAO photo)*

Left *A bundle of flax ready to be processed. (USDA photo)*

dollars per acre is invested in insecticides. When the crop is mature, defoliants* are applied to make the plants drop their leaves before the cotton is mechanically harvested.

The *seed hair*, which is the cotton fiber, is from one to six thousand times as long as it is thick. When the seed ripens, the hairs become flattened, twisted filaments containing about 90% cellulose. The high percentage of cellulose is responsible for the durability of cotton fabric, giving it the tensile strength necessary for the clothing and the myriad of other items woven from it.

Flax

The flax family comprises a dozen genera and 150 species widely distributed throughout tropical and subtropical zones. One variety of *Linum usitatissimum* is grown for its fiber. Others of the same species are cultivated for the seed, known as linseed, from which commercial linseed oil is extracted. Flax varieties grown for seed tend to be bushy and many-branched; those cultivated for fiber branch toward the top of the stem only.

Flax is highly sensitive to local weather conditions, which significantly affect the plant's growth more than the state of the soil. A temperate and even climate, free of heavy rains or frost, is necessary for optimal growth for fiber production. Moist winds are favorable. A hot, dry summer results in a short and rough but strong fiber; a moist growing period produces plants having a fine, silky but durable flax.

A difficult crop to grow successfully, flax is dependent on plentiful fertilizers, crop rotation, and well-timed harvesting. In Belgium, farmers will not grow flax on the same plot more than once in seven years, nor will they cultivate it on land that has been planted in beans, peas, potatoes, or clover during the previous year. Flax is susceptible to numerous diseases and to insect damage. Many hand operations are required during the growth of a crop, which increases the labor cost. While it can be a profitable crop when circumstances are favorable, flax is a less certain crop than many.

The flax plants are usually harvested when the stems are in transition from a green to a yellow color. The fiber will supposedly retain strength and its glossy, silk-like quality if collected at this stage. Experienced farmers generally harvest flax when the lower two-thirds of the stem has become yellow and the leaves have dropped. However, this varies according to the opinions of the growers. If the plants are left standing

---

*The same kinds of defoliants were used in Vietnam, and have been related to birth defects and other anomalies. Ecologically, they are indisputably a detriment.

too long, they will lignify and the fiber quality will be lessened. But if harvested too soon, the yield will be reduced.

Harvesting is usually accomplished by pulling the plant from the ground, to attain maximum fiber length and to avoid the mixing of field weeds with the crop. After being pulled, the stems are placed on the ground and left to dry for a day or two. Then the stems are arranged into shocks to complete the drying process, which may take another 12 to 14 days. Sometimes the plants are simply arranged in well-ventilated bundles, known as *barts*. When drying is completed, seed balls are removed by beating and separating the straw, a process known as *rippling*, customarily done by machine.

The next and most important step in the preparation of flax fiber is *retting*. The stem of a flax plant has 30 or so fiber bundles in it. In the process of retting, the flax stems are exposed to water, fungi, and bacteria, which decompose the materials enclosing the fibrous part of the stem but leave the fibers intact. The two main techniques involved are *dew retting* and *water retting*. In the former, the flax stems are laid out in fields, and turned once every week. The retting is accomplished through natural moisture. In the latter process, the stems are soaked in water—either in ditches, rivers, or vats.

For centuries flax was the most important fiber in Europe. It has been found in materials used to wrap Egyptian mummies, attesting to its antiquity as a cultural fiber. Today, it is grown primarily in temperate areas of Europe and in the USSR, which produces about half of the annual world yield of slightly less than a million tons. A minor industry in Japan and other parts of Asia and the Near East, flax cultivation has almost ceased in North America. Even so, few materials are better for making strong yet pliant fabrics.

Ramie

Ramie fiber comes from the stem of *Boehmeria nivea*, a kind of non-stinging nettle. There are about a hundred species of *Boehmeria* in all. However, despite experimentation with various others, *B. nivea* is the only one productive of a significant fiber. A native of China, the plant has been cultivated for hundreds of years, but was not introduced into the West until the early 1700's, when it was brought into Holland as an ornamental plant for botanical gardens.

Ramie yields a tough, supple fiber, but there are several cultivation and production problems. It is 16 times more exhausting to the soil than cotton, two and a half times more so than hemp, and six times more than flax. Also, retting is not very effective in removing the cementing vegetable gums which must be extracted before the fiber can be spun. Gum content can be as high as 45% or as low as 10%. Usually it is within the range of 20% to 30%. In China, the gum is removed by repeated washings, or treatments with wood ashes, after which the ramie fibers are washed and dried. In commercial degumming processes, chemicals are used; but too strong a chemical concentration will weaken the fibers. As the production of ramie fibers requires greater expense than other fibers, the plant's use is limited. It is grown in Louisiana and other areas of the southeastern United States, but its economic relevance is negligible.

Sisal

Sisal, a "hard" fiber, is obtained from the leaves of an agave plant, *Agave sisalana*. The fiber was first used in Central America and Mexico, where sisal was exploited during pre-Columbian times before the plant had ever been cultivated. (*Agave fourcroydes*, yielding the fiber *henequon*, is also grown in Mexico, but is of minor importance.)

With industrialization the binder-reaper resulted in an expanded outlet for sisal. The plant fibers were used chiefly to make binding twine. Sisal was introduced in Florida as a crop in the 1830's. It was not a success. Later experiments in Africa and the West Indies were successful. Today East Africa—Tanzania, Mozambique, and Kenya—supplies most of the sisal. Brazil, South Africa, Angola, Madagascar, and Haiti are other producers.

Successful sisal cultivation depends on suitable climate and soils. Although most agave grows wild in dry terrain, it needs abundant water for rapid growth. Most important, it can adapt to a wide variety of well-drained soils. Waterlogging and salinity will quickly kill sisal plants, and the soil must not be too acid or low in nutrients. Otherwise, the plant's tolerance is basically broad. Temperatures should be high and relatively even; the plants prefer full sunlight. Shade promotes weak, waxless leaves which have a low-quality fiber content.

A typical sisal plant produces 300 leaves during its lifespan; on flowering, the plant dies. Under optimal conditions, a plant may flower in as few as five years. Under less favorable circumstances, it may grow for 50 or more years before flowering, a characteristic to which it owes the popular name of "century plant." In commercial situations, leaves are usually ready for cutting in two and a half to three years. A leaf suitable for removal is in a nearly horizontal position, and almost at a right angle to the plant's base.

To harvest, the leaves are cut by hand; their spines are removed; and they are taken to a center for decorticating. After crushing, the pulpy tissues are scraped away mechani-

Above Sisal fibre is grown at
Tonj in southern Sudan. (FAO
photo)

Right Cannabis (hemp or
marijuana) is grown for its
fiber in many parts of the
world. (USDA photo)

cally, the loose tissue washed from the mass, and further cleaning is done by shaking, rubbing, and brushing the fibers —either by hand, or with the aid of special machines. The sisal fibers obtained in this way are principally sheathing cells around vascular strands; the vascular tissues are destroyed during the decorticating process. Waste materials from this procedure are a source of fertilizer nutrients. Fibers of too low quality for rope or twine are pressed into fiberboard.

Hemp

Hemp (*Cannabis sativa*) is one of the oldest cultivated plants in the world. Its origin appears to have been Central Asia. From there it was carried to China, where there are still more varieties than in any other area of the world. It has been cultivated in almost all temperate countries. *C. sativa* is commonly known as Italian hemp, since Italy once was the chief exporter. Currently, leading producers are the USSR, India, Yugoslavia, Hungary, and Rumania, approximately in that order. The annual commercial yield is not large, being about a third of a million tons.

Although there is only a single species of *Cannabis*, there are numerous varieties. The plants' fiber yield depends on the variety grown, and also the climate. Those plants cultivated for fiber have tall stalks, few branches, and few seeds. Plants grown for marijuana or hashish are shorter and many-branched (see Chapter 16).

Hemp cultivated for fiber is usually grown in temperate regions. Among the fiber-producing varieties, some grow better in more southerly areas; others are more suited to cooler temperatures. The latter require relatively low temperatures for growth, their vegetative period is short, and they are, as a result, shorter plants. Hemp varieties adapted to more southern climes do well in high temperatures, need a long growing period, are generally taller, and mature later.

The plants may be from 3 to 14 feet tall, varying according to climate and variety. If grown close together, the stalks will bear few branches; so, for fiber production, seeds are sown in close proximity to one another, as the stem is the important plant part. Hemp is dioecious, having staminate and pistillate flowers on different plants. Harvesting is done when the staminate plants are flowering.

The hemp is cut by hand with a hemp knife, which resembles a small sickle. Plants are cut off only 2 or 3 cm from the ground. The stems are left on the ground for dew retting, or put into ponds, or sometimes steam retted, (although this latter process, while faster, adversely affects the quality of fibers). After retting, the core and bark may be shaken away by

Above *Harvesting of jute in Nepal. (FAO photo)*

Left *The woody core of the jute is removed by pulling the stems backward and forward through the water. This is the last stage of the retting process. (FAO photo)*

hand. The fibers are twisted into yarn or twine on hand spindles; then woven into a coarse fabric, or used in the production of rope. Hemp fabrics are by tradition used for clothing corpses and as mourning garments in the East. Other uses are as nets, sail-cloth, canvas, and tarpaulins. The primary use of hemp today is as a substitute for flax, which it resembles although hemp is coarser and less flexible.

Other Fibers
Other fibers include jute (genus *Corchorus*) a bast fiber which is poor in quality, being rough and weak, but which is grown in quantity because of its low cost. Pakistan and India collectively grow more than 90% of the annual yield of three million tons. Manila hemp, or *abaca* (*Musa textilis*), is a relative of the banana. Grown predominantly in the Philippines, abaca produces moisture-resistant lightweight fibers with high tensile strength which are made into ropes, fishing nets, basketry, and miscellaneous items for export. Brush and filling fibers, of minor economic importance, may range from the "silk-cotton" hairs of tropical kapok trees (*Ceiba pentandra*) to shredded bamboo. Coconut fibers, or *coir*, are useful as filling and padding material or for making mats. Pineapple plants (*Ananus comosus*) yield a fiber suitable for "Piña cloth," rope, twine, and fishing nets. Fibers taken from the seeds of wild milkweeds (*Calotropis*) in India constitute *akund floss*, which may be used as a stuffing for mattresses or cushions, and as a substitute for kapok, although it is inferior in quality.

Wood
Wood, generally speaking, is the tough, fibrous supportive and water-conducting substance beneath the bark of trees and shrubs. As cells in the apical meristem of a wood-producing plant divide and elongate, tissues are laid down; one is a thin cylinder of secondary meristem, the *cambium*, located between the bark and the *xylem*. Xylem cells next to the cambium remain physiologically active, and are referred to as *sapwood*. When the sapwood dies it is permeated with tannins and dyes—becoming heartwood, the material that we think of as wood. Functionally, except for some cells in the sapwood, little of a tree trunk is "alive." The wood is dead, and only the cambial region—and some of its recent derivatives, such as phloem—is composed of living cells. The growth of heartwood is seasonal, even in the tropics. In temperate forests it is related to winter-summer cycles, and in the tropics it is a function of wet and dry seasons.

The "rings" of wood are a useful indicator of growth patterns of a tree, and may serve as indicators of weather changes in paleoclimatological studies. Relating ring patterns to ancient climates is the focus of *dendrochronology,* the science of dating past events by the study of growth rings in trees. Paleontologists may employ this technique in attempts to explain sudden disappearances of a species, for rapid extinction can be due to climatic changes with which an organism cannot cope. Archeologists, too, may extract information regarding changing settlement patterns, the collapse of civilizations, and other data related to possible weather changes.

A principal use of wood is for construction. Typically, it is a good insulator with numerous cellular air pockets present, although some woods—particularly tropical ones, such as ebony—may be so dense as to exclude most trapped air. The natural durability of wood to fungal and insect attacks, an important factor when considering its use in building, depends on the amount of tannins, phenols, and related compounds that are deposited in it. If kept dry, most woods can be preserved indefinitely. Many species of tree woods are naturally resistant to decay, because of compounds evolved to protect the living trees against insect and fungal attacks. Notable among these resistant trees are redwoods, cedars, and walnut. Commercially, some woods (telephone poles) are frequently given additional protection by treatment with creosote or phenols.

The *hardness* of wood depends on the quantity of lignin deposited. Lignin may comprise 20% to 35% of the heartwood. Another factor determining the hardness of wood is how easily the cells can be separated when mature. Oak, elm, hickory, maple, and ash are relatively hard temperate woods, though none of these are as hard as certain tropical hardwoods, such as mahogany.

The less lignin in wood, the more the wood will shrink on drying, with the accompanying risk that items manufactured from freshly cut wood may warp, twist, or split. With softer woods, shrinkage can be as much as 20%. Quality lumber is usually kiln dried to lessen shrinkage.

Currently, most of the wood used for commercial purposes is from softwood species, particularly pine. In the United States, for example, 80% of forest land is planted in pines because of their rapidity of growth. Hardwoods are usually imported from tropical countries, where the trees are simply cut out of jungle areas. Because of the time required for the trees to mature, commercial growing is generally not considered. However, some hardwoods are cultivated.

In tropical areas the hardwoods are felled and left to lie for about six months during the dry season; then the trees are floated down rivers to the nearest mill where they are cut and shipped as raw lumber. The most intensive management of hardwood trees is practiced in Malaysia, where teak is commercially important. Customarily, acreage is cleared and farmed for two years, to take advantage of the fertile soil. Then teak seedlings are transplanted from nurseries. Food crops are often cultivated between the growing trees until the trees are large and provide too much shade.

Teak will yield lumber in about ten years when grown in this way. Some mahogany is also farmed in Malaysia, in similar fashion. But in most tropical countries the attitude that forests are merely storehouses to be exploited still predominates, and there is little or no management of forest resources. Most of the so-called Philippine mahogany comes, in reality, from Borneo, where it is cut out of the jungles by Australian lumberjacks.

Cellulose is produced from and is utilized for various industrial purposes such as paper pulp. The variation in annual consumption from one country to another is great, with the people of the United States having a large lead, using an average of 400 pounds per person. In Britain the figure is 164 pounds per person, in France 84, in Japan 47, in the USSR 25, and in China a typical individual's share is only 2 pounds!

Modern Forestry Today, *forestry* has assumed the proportions of big business. In the U.S. and Europe, *silviculture*—the cultivation of timber —has largely replaced the less efficient approach of logging wild trees. Small-scale forestry practiced on private farms is certain to diminish, as lumbering becomes increasingly the business of corporations. On farms, the trees are of secondary importance to agricultural crops. Usually maximal tree growth cannot be attained, and often farmers are not aware of the value of the trees.

Modern forestry corporations consist of a network of interests, ranging from silviculture to the production of lumbering products, from paper to plywood to packaging materials. Since forest products account for more than 33 billion dollars of the world's annual export total—about a sixth of that for all food and feeds, and about 3% of all world trade—forestry is understandably an attractive business. It is of most importance in Scandinavian countries, Canada, and parts of the United States. South America and Asia export relatively little in the way of forestry products, and in Africa the exports are not great, although they are appreciable in relation to the total economy.

In contemporary silviculture, the direct seeding of an area is usually the most economical means for reseeding a forest. This can be accomplished by leaving seed-bearing trees intact or by transplanting seedlings begun in a nursery. With increasing demands for lumber, it is important to achieve maximum efficiency in silviculture. Studies of energy entrapment by various forest ecosystems should lead to improved productivity, as will the seed improvement of tree strains.

In the past, lumbering was accomplished by lumberjacks, rugged individuals who lived a harsh existence felling trees by hand. Today, lumbering is mechanized, but the major effort and expense is still centered on cutting the timber and transporting it to market. In modern logging operations, the trees are felled with power saws, and huge machines with giant shears. Another machine lifts the fallen tree, sending it through a tunnel of blades and saws that delimb it, cut the trunk into segments, and even remove the bark. Other massive machines are employed to lift the cut logs onto semitrailer trucks, on which they are carried to sawmills, paper processing plants, or other points where they are converted into usable products. The machinery needed for these sophisticated lumbering operations is expensive; some costing several hundred thousand dollars or more, a factor which excludes the smaller operator. However, the machines are economically feasible investments. Whereas a single lumberjack can harvest one cord of wood during the course of a day, an individual assisted by machinery can harvest seven or more cords.

In the tropics, forestry techniques are generally undeveloped, with the exception of extensive teak plantations in the East. The custom of cutting hardwood trees out of jungle areas has depleted the natural supplies of some hardwoods, such as mahogany. Climate also presents a problem to the tropical silviculturist. Rain, heat, and humidity may cause rapid deterioration of expensive machinery, and insects can severely damage seedling trees. Also, transportation to commercial centers from growing sites may be undependable or nonexistent. For these reasons, tropical logging operations are likely to continue to depend on wild trees, rather than risk the investment of immense capital outlay in what may be a chancy venture. This situation is regrettable because the wild populations of some valuable tropical hardwoods are being so thoroughly decimated that seed stock for future replenishment may prove difficult or impossible to obtain.

It is probable that, with increasing world demands for more food, land now devoted to forestry will be redirected to food production. It has been estimated that in the United States, more than a hundred thousand acres of forest land will soon be converted to other uses—leaving about a half-billion acres in timber. This may seem a large area, but when one considers that the world population will continue to increase, and with population increase the increased demand for forestry products, it becomes obvious that lumber and related materials will be more expensive in the future. The use of particle board and other techniques to make better use of all parts of the tree will contribute somewhat to alleviating the problem of lumber shortage.

Plywood, made from layers of wood glued tightly together, has achieved a rising popularity during the past few decades due to several factors. Improved resin glues are now stronger than the wood itself, making plywood a hardy material for marine and exterior construction. Also, the use of plywood in construction reduces the time involved, as it comes in standard 4' × 8' pieces that can be rapidly fabricated. In 1950, 2 billion square feet of plywood was manufactured yearly. By 1970 the total had increased to 10 billion square feet, an indication of the increasing economic significance of plywood.

Wood Products **Paper.** Paper is essentially a mat of fibers. The first process for making paper products was originated by the Chinese during the first century AD. Stems, silk, and bark were shredded together, dispersed in water, sieved onto a mat, and dried. The contemporary process is still much the same in principle. Fibers are separated and dispersed, matted, then filled with clay or other substances to add weight and body, and finally coated to make them more receptive to inks and other printing matter.

In Europe during the Middle Ages, parchment, made from animal hides was used frequently as paper. Paper was also made from old linen rags. Following the invention of the printing press, the demand for paper rose and wood fibers came into use. However, a process for shredding wood into fibers was not developed until 1850. Today 90% of all paper manufactured is made from wood fibers. The remaining 10% is made from flax (as in money and cigarette papers), cotton fibers, and other materials.

When wood is used to make paper, the logs are sprayed with water and ground to break up the cellulose fibers. In the manufacture of newsprint, lignins and other organics are not removed, which accounts for the yellow discoloration and brittleness characterizing older newspapers. For the production of higher quality paper the lignin must be dissolved, a process that significantly increases the production cost. To dissolve the lignin, the raw materials must be soaked in bisulfite at 130°C to 150°C; the quantity of lignin extracted is dependent on the time the material is cooked. In the *sulfate* or

Above *Charcoal development work in Uganda. Kiln operator fitting smoke stacks to the charcoal kiln. Each kiln can convert about 100m3 of waste vegetation to approximately 6000 kilograms of charcoal a month. (FAO photo)*

Left *Fast-growing eucalypti in Australian forests provide wood for a wide range of domestic uses from hardwood for building to pulp for cardboard cartons. Eighty-eight percent of all domestic wood used in Australia is eucalyptus. Seeds are much in demand by other countries needing quick-producing stands.*

kraft process, wood chips are digested in a solution of sodium hydroxide and sodium sulfide at 170°C to 180°C. This method of dissolving the lignin is especially effective with resinous pinewood. The *soda process* employs sodium hydroxide, and is chiefly an agent in the fabrication of book and magazine paper.

Charcoal.   The cutting of trees for *charcoal,* a black form of carbon produced by partially burning wood in containers from which air has been excluded, has been an element of rural human cultures for millennia. In our contemporary, densely-populated world the environmental impact of this practice is increasingly negative as so much wood is required for a relatively small amount of energy. Unfortunately, government restrictions against cutting trees would effect intense hardship on many groups of poverty-level, rural people. Yet, the deforestation due to charcoal production is so extensive that it cannot be ignored.

The adverse effects of deforestation as a result of charcoal burning may be illustrated by the example of Mexico. In that country the practice of charcoal burning is first evident from archeological data more than 2000 years old. The abandonment of the Teotihuacan civilization in the Valley of Mexico was probably due, in part or wholly, to crop failure. The crop failure may have arisen, at least partially, from the drying up of streams and from the lessened rainfall accompanying the destruction of large tracts of forest. At Teotihuacan, lime cement covered almost all buildings and comprised the pavement. Since this material is extracted by burning limestone, presumably a considerable quantity of charcoal was needed. To supply the charcoal, the forests were cut down—with the result that an entire civilization's food supply was affected.

Today, Indians and rural *mestizos* in Mexico must burn ten times the quantity of wood as the limestone being reduced. Charcoal also remains an important component of their cooking fires. Since they have metal axes at their disposal, cutting large amounts of wood is easier for them than it was for their pre-Columbian ancestors.

Trees used in making charcoal are frequently hardwoods, as the greater density of such woods is conducive to longer burning. Very dry wood is inferior to that which is freshly cut; the drier wood burns too quickly. However, in an effort to slow deforestation, the Mexican government enacted legislation outlawing the felling of live trees for charcoal. The result was that roots of living trees were surreptitiously cut to hasten the trees' death and make them legally eligible for charcoal production. The law was enacted primarily to limit the activities of large-scale manufacturers. Enforcement on a local, village level is virtually nonexistent.

To make charcoal, trees are cut into small pieces, which are graded by size, and then set into holes in the ground, in cone-shaped stacks made by placing the larger pieces on the bottom. A central core is arranged in the stack, of highly combustible pine and *ocote,* (veins of resin cut from the pine). Then the cone is covered by an 18-inch layer of dirt, and the pine is ignited. New wood is fed in as needed to keep the stack burning. Smaller pieces usually become charcoal within about 36 hours; larger ones may take five days or more.

The fresh charcoal is removed and cooled and then stored in a dry place. Besides having local uses, charcoal production is a means for supplementing the *campesinos'* meagre incomes. A bag of charcoal weighing 35 kilos (about 77 pounds) will bring fifteen pesos or so ($1.20) when sold to dealers or directly to the inhabitants of towns.

There is no easy answer to the problems associated with extensive charcoal consumption. The need for increasing energy sources poses a dilemma to all societies today. We cannot expect the rural peoples who depend on charcoal for a variety of uses to abandon it as an energy source unless an equally inexpensive substitute is provided. Since the cost of energy is increasing globally, feasible substitutes may not be forthcoming. However, it is clear that continued deforestation will deplete the trees necessary for charcoal, while at the same time causing environmental disruption which may have tragic effects on local agriculture.

Tobacco

Tobacco cultivation is complicated, as the plants require considerable care, and are subject to a variety of diseases. Because tobacco is grown in diverse areas, localized techniques for its cultivation have evolved, but some general principles are applicable anywhere. Due to the small size of the seeds and the seedlings' inability to compete with other plants, seedlings are almost always begun in seedbeds, and then transplanted later to fields. The seedbeds are customarily sterilized, often with chemicals such as methyl bromide. The plants are susceptible to diseases and parasites (especially nematodes), and disease control and treatment of the soil are imperative. The seeds are small in size, and are usually mixed with some other material, such as sand, ashes, or even cornmeal, before being spread over the seedbed. An ounce of seeds is adequate for 2,000 square feet of seedbed planted at a density of 60 plants per square foot. The seedlings may be covered with a cloth for shade, or with polyetheylene plastic, which provides better protection and humidity. Smoking is not allowed in the vicinity of the plants, as diseases are easily introduced. Hybrid seeds, resulting from the hand-pollination of a male-sterile line, produce a more uniform, higher-yielding and rapidly-germinating crop, for which there is growing demand.

Tobacco requires a clean-cultivated soil, which is a condition conducive to erosion, such as occurred on the early tobacco plantations in Virginia. Thus sloping land is avoided. Abundant fertilizer is required; the soil should be rich in potash and phosphorus, but containing low levels of nitrogen. Seedlings are usually transplanted to fields in ridges three feet apart, with the plants two feet apart in each row, resulting in 7000 plants per acre. A general practice during growth is *topping*, the removal of the terminal growing tip, to prevent flowering and seeding and the accompanying loss of nutrients that would detract from leaf-making. As a corollary procedure, axillary shoots are removed as necessary for the production of fewer, larger leaves.

The tobacco leaves are harvested from the bottom upward on each stalk, at the rate of two or three leaves at a time, as they reach desired levels of maturity. The leaves must be handled gently to avoid breakage. After cutting, they are tied into bunches and hung in large curing barns (although Turkish tobaccos may be cured in the sun) which are heated by flues from wood or coal fires. Temperatures in the curing barns vary according to the locale and the kind of tobacco. Curing may begin with a temperature of 29°C (85°F) on the first day, increased to 32°C (90°F) until the leaf is well-colored. After this, the temperature is raised to 38°C (100°F) until all green color is gone from the leaf. After curing, which takes only a few days, the leaves are allowed to absorb enough moisture to give them pliability necessary for handling.

The classification of tobacco types is arbitrary, and is based on the given strain, method of cultivation and curing, color, the geographical area of cultivation, and the particular use for which the tobacco is intended. The United States Department of Agriculture has designed a classification system consisting of classes, types, and grades: flue-cured, fire-cured, air-cured, cigar-filler, cigar binder, cigar wrapper, and miscellaneous. Such a classification system is a matter of convenience, not of intrinsic distinctions.

The tobacco plant is a glandular down-covered perennial that is commercially grown as an annual. It is coarse, with large, alternate leaves and a terminal cluster of five-parted, tubular flowers. The leaf, which is the economically valuable part of the plant, has a strong, harsh flavor and odor until it is fermented, a curing procedure that imparts the customary aroma to tobacco.

A characteristic component of tobacco, *nicotine* ($C_{10}H_{14}N_2$), is an alkaloid which volatizes readily with heat, producing a somewhat strong vapor. As a contact poison, it is lethal for most insects. There is strong evidence that it is carcinogenic in human beings. *Nornicotine* ($C_9H_{12}N_2$), another alkaloid, is more abundant in some strains than is nicotine. The smoke of any tobacco product contains the gases oxygen, nitrogen, and carbon dioxide, as well as smaller quantities of carbon monoxide, hydrogen sulfide, hydrocyanic acid, and ammonia. Solids and liquids comprising the "tar" and present in the smoke are: nicotine and various derivatives; pyridine compounds; resins; essential oils; methyl alcohol; acetone; formic, butyric, and acetic acids; and phenols. The nicotine is of most interest to human beings, as it is responsible for tobacco's mildly narcotic qualities.

Suggested Readings

Carter, G. L., and Horton, Paul M. *Ramie*. Baton Rouge, Louisiana: Louisiana State University Press, 1936.

Duerr, W. A. *Fundamentals of Forest Economics*. New York: McGraw-Hill, 1966.

Herndon, Melvin G. *William Tatham and the Culture of Tobacco*. Coral Gables, Florida: University of Miami Press, 1969.

Kirby, R. H. *Vegetable Fibers*. London: Leonard Hill Books, Ltd., 1963.

Lock, G. W. *Sisal*. 2nd ed. London: Spottiswoode, Ballantyne, and Co., Ltd., 1969.

Janick, J.; Ruttan, W.; Schery, R. W.; and Woods, Frank W. *Plant Science*. San Francisco: W. H. Freeman and Co., 1969.

Schery, Robert W. *Plants for Man*. 2nd ed. Englewood Cliffs, New Jersey: Prentice-Hall, Inc., 1972.

Stoddard, Charles H. *Essentials of Forestry Practice*. 2nd ed. New York: Ronald Press, 1968.

U.S. Department of Agriculture. *Tobacco In The United States*. Misc. Pub. 867, USDA, 1966.

CHAPTER

8

FRUITS, ROOTS, AND VEGETABLES

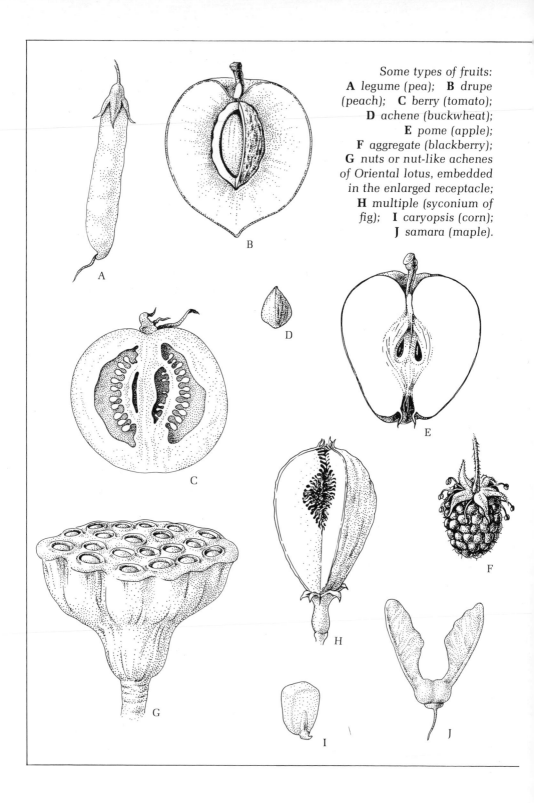

Some types of fruits:
A legume (pea); **B** drupe
(peach); **C** berry (tomato);
D achene (buckwheat);
E pome (apple);
F aggregate (blackberry);
G nuts or nut-like achenes
of Oriental lotus, embedded
in the enlarged receptacle;
H multiple (syconium of
fig); **I** caryopsis (corn);
J samara (maple).

The categories comprising fruits and vegetables tend to be cultural rather than botanical. Thus the two terms are somewhat arbitrary, a "fruit" in one area may be a "vegetable" in another. By botanical definition, a fruit is a matured ovary. Edible plant parts which are, in fact, fruits by the strict definition, but which are frequently thought of as vegetables include tomatoes, peppers, egg plants, squashes, pumpkins, cucumbers, and many others.

Fruits

A variety of fruits have an established position in national and international commerce, such as oranges, bananas, and pineapples. However, most are consumed at a relatively local level. Many excellent fruits remain exotic except within limited areas. For example, litchi nuts, a fruit grown primarily in China, eastern Asia, and in parts of Florida, and pitahayas and tuna, the fruit of certain cacti, are seldom items of general commerce. Yet numerous fruits which are now common were rarities only a few generations ago. Today, due to rapid transportation, ripening inhibitors, and improved storage techniques, they are found frequently on our grocery store shelves. An orange or banana in a Christmas stocking may have been an unusual treat when our grandparents were young.

The diversity of fruits and vegetables available to us in most American supermarkets today results from the relative ease with which perishable crops are moved from one part of the country to another, and the refined preservation techniques developed recently. Previous to refrigeration facilities, most fruits were necessarily local commodities due to the rapidity with which they spoiled. Until recently, this disadvantage was overcome partially by drying, sugar preservation (candying), treatment with metabolic inhibiting chemicals or growth retarding hormones, canning, and freezing.

Most fruits are more popular than their nutritional content might warrant, particularly from the standpoint of protein content. An average fruit is 80% to 90% water and contains from 1% to 4% protein. Of course, human metabolism requires more than protein, and fruits do tend to be high in some trace metals and vitamins, especially vitamin C. A few fleshy fruits do have considerable protein content, as much as some grains. When this occurs, a fruit may be a dietary staple. Coconut, for example, is about 10% protein.

An interesting aspect of the relationship between fruits and human beings is that, until quite recently, fleshy fruits were relatively unchanged in comparison to their ancestral types. (As we have seen, this is not the case with many grains.) There

Above The papaya, a New World plant domesticate, has spread throughout the world's tropical zones. The plant is productive and the fruit has excellent nutritional qualities. (FAO photo)

Left Avocados are one of the more important fruits originating in the New World. (USDA photo)

have been seedless pineapples and bananas since the beginnings of recorded human history. Today, most of the characteristics of commercially available fruits—such as large size as a consequence of polyploidy and seedlessness in grapes, oranges, and tomatoes—are the products of research in plant genetics and horticulture over the past several decades.

Since fruits have been local in production and consumption during most of human history, we believe that it is most meaningful to consider them in relation to their geographical origins. These origins include the New World, the Mediterranean, the Indian subcontinent, Africa, and Southeast Asia.

Papayas

The papaya (*Carica papaya*) was for some time thought to have originated in India, but recent investigations have established that the fruit is of New World origin, probably from the lowland Caribbean coast of Mexico. It was a food in the Mayan diet; and when commerce started between the New and Old Worlds, it was carried to Africa, India, and the Caribbean islands. Today, many types of papayas are grown in most tropical and subtropical regions of the world. The soft, subtly-flavored flesh ranges from yellow, to orange, to a dark red-orange. Picked for shipping when they are green, papayas have their best flavor when the outer skin has turned yellow. Fruits range in size, according to the variety and the growing conditions, from less than a pound to more than 10 pounds each.

Papayas are low in calories, and contain more vitamin C than apples or citrus fruits, more vitamin A than carrots, and also appreciable amounts of vitamins B_1 and B_2. They also have an abundance of *papain*, a proteolytic enzyme capable of digesting animal proteins, which is obtained from the milky juice of green fruits. This enzyme aids human beings in the digestion of other foods, as its activity is intermediate between that of two vertebrate enzymes, pepsin and trypsin. Commercial meat tenderizers are among the more economically relevant products obtained from papayas. Papain is also used by pharmaceutical companies in various preparations. Throughout the tropics, papaya is considered an aid for individuals having digestive problems. Papaya powder is a component of some toothpastes, and is effective in treating the gum disease, pyorrhea. Papain has been used as a beer clarifier for years—causing a smoother, better-flavored beer.

Avocados

Avocados (*Persea americana*) are native to southern Mexico and Central America, where the name of the fruit is *aquacate*.

They are sometimes referred to as "alligator pears" because of their rough, green or black skin. Contemporary avocado fruits have a thick, soft green flesh enclosing a large seed. The archeological record indicates that their size was increased considerably through selection during prehistoric times. Nutritionally, they are significant because of a high fat content (up to 30%). Seventy-eight percent of the fat is unsaturated. Eleven vitamins are present, including A, C, B_2, B_6, E, and K, and 14 basic minerals. Avocados are an important dietary element in many tropical regions, although the protein content is low (only about 2%).

In nature avocado fruits grow on evergreen trees which attain a height of 50 feet. Commercial trees are smaller. The fruit is hard while still on the tree but softens rapidly after picking. In the United States, avocados are cultivated commercially in southern California and Florida, and are shipped throughout the country.

Pineapples

Pineapples (*Ananas comosus*) are a tropical fruit native to South America. They have long been cultivated, and were a crop of the Peruvian Incas and their ancestors. By the time of Columbus' arrival in the New World, pineapples had spread thoughout Central America and Mexico, and had been carried to the West Indies, where they were first encountered by Columbus in 1493 on the island of Guadalupe.

The pineapple is a multiple seedless fruit—a product of the fusion of many fleshy flowers on a common axis. Although the stem of a plant bearing fruit will not produce a second fruit, after 10 to 18 months another flowering stem arises from near the plant's root. Thus a single plant may be productive for 10 or more years. Pineapples grow best in volcanic, well-drained soil. They propagate by suckers or shoots arising from the parent plant. Sometimes growers cut away the leafy portion of the fruit's top and plant it. Today, the principal growing regions for international markets are Mexico, Hawaii, and South America, particularly Ecuador.

Strawberries

Composed of several species, this fruit is derived from New World plants. The genus *Fragaria* is now grown everywhere in the northern hemisphere. Currently one of the largest fruit crops in the U. S., Europe, and Japan, strawberries were originally brought from North America to France in 1712. As the fruits became popular, strawberries were grown in more and more areas. In Europe, the North American varieties were crossed with the smaller varieties indigenous to that continent.

Strawberries are not true berries (see Chapter 2), but an aggregate fruit consisting of separate tiny pistils (akenes) on a fleshy receptacle. They are perennial herbs, propagating themselves by stolons. Modern strawberry cultivars deserve particular attention. These are truly genetically designed fruits in which there has been little chance selection. The fruits tend to be expensive in the market largely because they must be picked by hand, a time-consuming process that adds to their cost. The bulk of the commercial crop becomes the ingredients for jams, jellies, and other preserves.

Tomatoes

The tomato (*Lycopersicon esculentum*) is one of the principal fruits, economically, of a large family (*Solanaceae*) of New World plants which includes potatoes, ground cherries, peppers, and tobacco. Small-fruited tomatoes are found throughout the Americas, but the large-fruited varieties probably originated in Peru and Ecuador. Tomatoes have been an important fruit in local diets of the Neotropics since before the European conquest. Plants were supposedly carried to Europe from the New World in 1523. By the 17th century, tomatoes were grown throughout the European continent, but chiefly as a garden ornamental rather than as a food plant. Acceptance in Europe as a food was initiated in Italy and Spain. Today, tomatoes seem almost synonymous with native Italian foods. In Central American markets, one can find various types of tomato cultivars, most grown in small family gardens.

A tomato is a fleshy annual, best suited to warm growing conditions. It will not set fruit if night temperatures drop below 64°F, as the fertilized flowers will drop from the plant. In hot, moist climates yields are usually good. In the tropics, three or more crops are often produced every year. In the United States, tomatoes are grown commercially in Florida, California and a few more northerly locations. Much of the total annual crop is canned, or becomes the base of catsup or tomato paste. When canned, tomatoes retain most of their vitamin C. Greenhouse cultivation during the winter is also profitable, as tomatoes are particularly adaptable to hydroponic conditions, in which a nutrient solution containing essential nurturing elements is used as a soil substitute. Under such conditions, the tomatoes are often grown in elevated troughs.

Citrus

The worldwide popularity of *citrus* as a cultivated fruit is interesting because, from a nutritional standpoint, citrus is inferior to many other fruits. A typical citrus fruit is about 90% water, 5% to 8% sugar, 1% to 2% pectin, citric acid,

Above *Strawberry cultivation in winter in Japan. Japanese strawberries are oblong and grow to a length of about 6 cm. A major crop, the fruits are famed for their juicy, distinctive flavor. The special "Ishigaki cultivation method" has been adopted in Japan, in which concrete blocks are laid along sunny slopes and the strawberry seedlings planted in the soil between them. The concrete blocks, heated by the sun, help to hasten the growth of the plants. The entire bed is covered with vinyl sheets at night to prevent any damage from frost. A large quantity of these high-quality Japanese strawberries is now exported to many countries. (FAO photo)*

Left *Mangos, of East Indian origin, have become an important commercial crop in Africa, Mexico, and Central America. They are grown to a limited extent in south Florida. (USDA photo)*

protein, essential oils, and minerals. Other fruits surpass citrus in protein and in vitamin C content, supposedly their major nutritional virtue. On the other hand, most citrus varieties are relatively prolific; the fruits keep well and ordinarily can be transported without significant damage. They are attractive to consumers, having both appealing flavors and appearances. Some of the more popular commercial types, such as tangelos, are of recent origin, the results of hybridization.

Citrus trees have been part of a curious cycle in southern California and in Florida, the two principal areas of cultivation in the United States. When land was inexpensive, thousands of acres were planted in citrus, because the balmy climates of these two regions were favorable to their growth. Yet, within only a few decades of the spread of citrus orchards in these centers, the urban and suburban population explosion reclaimed much of the land planted in citrus. This is an ongoing development. Every year, more acreage of former citrus groves is covered with subdivision housing.

Citrus fruits include oranges, lemons, grapefruits, citrons, pummelos, sour oranges, calamondins, kumquats, mandarin oranges, and tangerines. Most of these are of Indian and southeast Asian origin. Only citrons were introduced into the Mediterranean before the Christian era, followed centuries later by lemons, limes, and oranges. After the decline of the Roman Empire, these spread to Italy, North Africa, and Spain, largely as a consequence of Arab influence in the Mediterranean during the Middle Ages. Not until the 15th century did the sweet orange reach Europe, probably as a result of Portuguese contacts with India and Malaysia. Citrus was later brought to the New World by Spaniards. Until recently, Florida dominated citrus production in the United States, and was the center of hybridization and other experimentation. In 1862, the seedless grapefruit was "discovered" as a mutant of the seeded variety and was rapidly propagated. Pink grapefruit arose in like fashion as a bud sport.

Citrus was introduced into California during the 1700's. The trees were cultivated around missions. In 1871, navel oranges were brought into California from Brazil. But not until extensive irrigation could be practiced did citrus become a major crop in California. Today, some citrus is also grown in Arizona and in Texas, especially in the Rio Grande Valley. The trees are frost-hardy, but they cannot withstand prolonged freezing. Groves are usually propagated by grafting or budding onto good root stock. The primary uses of citrus are as fresh fruit, or as frozen or fresh juice. The residues obtained from juice processing plants are used as cattle feed in Texas and Florida, where many beef cattle are raised. Pectin and essential oils are also obtained from the peels.

Mangos

The mango (*Mangifera indica*), one of the oldest of cultivated fruits, has been grown in India and Malaysia for more than 4000 years, and it is still a common wild tree there. Mangos were brought to the New World in the 1700's by Portuguese traders, and into East Africa at about the same time. The large evergreen trees are attractive, spreading a broad shade area where they grow. In the 1800's, mango trees were taken into Mexico, which is currently a leading grower and processor of the fruit. Numerous groves were planted in south Florida in the 1880's; that region is still the major producer in the United States.

The trees must be entirely free from frost if they are to bear fruit. A dry season is preferable during flowering, followed by abundant rains when the fruit is setting. Wind may reduce a potential crop significantly by causing the pendulous, rather loosely-attached fruits to drop before ripening. Even very heavy rains may cause them to fall.

The mango fruit varies in size from that of a large plum to that of a very large pear. The fruit is fibrous in texture, often stringy. Some of the hybrid strains, such as the Hayden mango, are very meaty and less coarse than others. A single sizeable seed with many fibers attached is in the center. More than 500 varieties of mangos have been named and are grown throughout the world in relatively warm locales. Because of a peculiarity of ripening, mangos have become important as a transported commercial food only recently. When picked green, the fruit never ripens properly, and often has a "turpentine" flavor, which may be present even in tree-ripened individuals of some varieties. Also, if mangos are allowed to ripen on the tree for too long, the same taste may be evident. Thus, a mango should be picked when it is "just right," a stage that takes some practice to discern. Until modern transportation could speed ripe mangos to the marketplace, the fruit was not marketed on a large-scale because of these features.

Commercially, mangos are usually propagated from grafts or cuttings because some varieties will not breed true from seeds. The fruit is an important part of local diets in most tropical and subtropical areas of the world, being rich in vitamins A and C. In Mexico, a powder is made from mangos and processed as a beverage component. The production of chutney, a vinegar-sugar relish, consumes many mangos. Much of the fruit goes to waste on the ground, where processing and shipping facilities are still lacking in some areas.

Dates

The precise origin of the *date palm* (*Phoenix dactylifera*) is unknown, but it has probably been a cultigen for as long as the coconut. The trees were grown in Egypt and the Middle East before written history began. Second only to the coconut palm, date palms have multiple uses for human beings. The fruit is eaten fresh and some species are used in making wine and vinegar. The seeds or "stones" of the berry are ground into meal and fed to livestock, or roasted as a substitute for coffee. Oil is also expressed from the seeds. The tree itself furnishes many of the same materials for the manufacture of cordage, baskets, and housing materials, as does the coconut palm. Nutritionally, the fruits are rich in carbohydrates (about 70%), and have some protein (2%) and fat (2.5%). They are rich in vitamin A as well.

Southern California produces nearly all of the dates grown in the Western hemisphere. Dates were introduced into the Coachella Valley of southern California in the early 1900's. Now, more than 30 varieties are grown in the state. The date palms are usually propagated by suckers—side shoots from the base of the trunk. Flowering begins during the fourth year, at which time the female (pistillate) inflorescences are dusted by hand with flowers obtained from male (staminate) trees, of which fewer are needed. In modern orchards, the developing fruit may be protected from birds and other potential damage by paper bags. A single date palm may yield more than 100 pounds of fruit annually. The dates are hand-picked, and then sometimes steamed to foster better quality. Further ripening allows the sugar content to develop and astringent constituents to precipitate.

The so-called Chinese date (*Zizyphus jujube*) is cultivated to a limited extent in South Carolina and northern Florida's coastal regions. It is not a palm, but a plant of the family *Rhamnaceae* which varies from the size of a small shrub to that of a tree. When dried, the fruits are similar in texture and taste to dates.

Coconuts

The coconut palm (*Cocos nucifera*) is found throughout tropical areas of the world. Carried from one area to another by ocean currents, the seeds germinate on the beaches where they come to rest. Their origin as a cultigen is in tropical Asia and Malaysia. Today, they are one of the most useful palms—an economic staple in many humid, tropical regions, and the basis for local economies.

The coconut palm's uses are many. The "wood" of the

Above left *The natural method of sun drying coconut "meat" facilitates oil extraction at a later date. (FAO photo)*

Above right *Coconuts are essential to tropical economies. Copra is the dried meat. Coir, used to make ropes and matting, is the fiber processed from the outer husk. Coconut oil extracted from the meat has multiple uses. (FAO photo)*

Left *Coconuts in the tree. (FAO photo)*

trunk is employed in the construction of houses. The fibers inside the husks are manufactured into yarn and woven into a coarse cloth. A variety of utensils are made from the seed coats. The leaves are used in basketry, and for roof thatching and wall structures. The "milk," a liquid endosperm within the seed, and the solid endosperm meat are nutritious foods. The tree's sap may be fermented into a toddy.

Coconut trees can be propagated only by seeds; they cannot be started from grafts or cuttings because they have no cambium or lateral meristem. Trees begin to bear in six to eight years, and thereafter will yield a steady crop for up to fifty years, under favorable circumstances. The fruit of commerce is from three to eight or more inches long, encased in a husk of fibers enclosed in a tough outer covering. The meat of the seed is encased in a hard, brittle shell which also contains the central watery endosperm (milk). Coconut milk, which has a distinctive and pleasant flavor, has a high content of amino acids. The meat may be eaten fresh. When dried, it keeps well, and constitutes a nutritional food source of 10% protein and 25% oil. The dried meat, known as copra, contributes significantly to the economies of many rural, tropical settlements.

In regions where the coconut is a staple, oil is extracted from the meat by shredding or grating, then boiling in water until the oil separates and rises to the surface. It is used in cooking, and was the sole source of such oil for the natives of Micronesia, prior to the introduction of other oil substances by Europeans. In young coconuts the meat is jelly-like, before the meat and milk separate. At this stage of development, it is a staple food for children in much of Malaysia.

Bananas

Greeks, Latins, and Romans mentioned bananas (Musa spp.) in their writings, referring to them as an Indian fruit. Polynesians distributed bananas throughout the Pacific, and Arabian merchants carried them to many parts of Africa. The common name "banana" originated in Sierra Leone. Portuguese and Spanish colonialists completed the fruit's introduction into most warm areas of the world. Even in the late 1800's there were some banana trees in the United States. De Candolle, author of The Origin of Cultivated Plants, wrote that bananas had "a diffusion contemporary with or even anterior to that of the human race" (note that he was equating "human race" with Europeans).

Bananas cultivated for food are of the species M. paradisiaca sapientum, M. sapientum, or M. cavendishii, according to which nomenclature one accepts. Although banana trees may grow quite tall, reaching a height of more than 30 feet, they are classified botanically as a perennial herb, lacking

Above Like most tropical
countries, the Ivory Coast has a
dry season and a rainy season
during which water is wasted.
Irrigated banana groves produce
four to five times more fruit than
plantations under dry cultivation.
The leaves remain green year
round, the bananas grow bigger,
and the bunches heavier. (FAO
photo)

Left Figs, an early domesticate of
the eastern Mediterranean, grow
in many hybrid varieties in
subtropical regions. (USDA
photo)

woody tissues. As fruit, bananas are predominantly sterile and seedless. Cultigens are propagated by rhizomes, or underground stems. These produce a cluster of large leaves and in less than a year a terminal inflorescence, the stem dying upon fruiting. Bananas are a popular lowland tropical crop. They are well adapted to the humid conditions of such areas, and will yield 10 or more tons of fruit to the acre with a minimum of grower attention. They are perhaps the most prolific of food crops when food production is measured against amount of land cultivated. Four thousand pounds of bananas can be grown in a space that would yield about 33 pounds of wheat, or 98 pounds of potatoes.

Bananas grow well on land from which rain forest has been recently hewn. The remaining forest encircling the plantations is an aid in buffering against winds which can seriously damage the plants' broad, delicate leaves. Well-drained soils are favored, although banana requirements are flexible. They are subject to diseases, especially "Panama disease," which results from a soil borne fungus, and the "Bunchy Top disease." They are also vulnerable to attacks by *sigatoka*, a leaf disease due to an air-borne fungus which causes irregular ripening and a reduction of the fruit, eventually killing the plant.

At the turn of the century bananas began to assume major economic importance in Central America. Panama, Costa Rica, Nicaragua, El Salvador, Honduras, and Guatemala were called the "banana republics" because their economies relied so heavily on the banana plantations, owned predominantly by American corporations. The holdings frequently included whole villages and the railways and other transportation connecting the plantations with ports.

Most of the bananas sold in the United States prior to 1964 were of the "Gros Michel" cultivar, also known as the "Poya variety," after Jean François Pouyata, a Jamaican who accidentally found the banana type in a plantation in Martinique in 1836. The large fruit of this variety was the only one seen in North American markets until the 1960's. Other varieties, such as the small lady-finger bananas, are considerably more flavorful. And plantains, starchy bananas that must be cooked, are much larger. However, for extensive international commerce the former are too perishable, and the latter too flavorless and bitter, unless cooked. So the Gros Michel remained the staple. Its disadvantage is a vulnerability to the diseases mentioned. Agronomists have introduced new disease-resistant varieties as substitutes, the most notable of these being the Lacatan and Valery.

The annual commercial yield of bananas is about 23 million tons, a figure that would be larger were it possible to include the quantity consumed on a local level. In humid tropical regions most households maintain a few banana trees, both for ornamental and dietary reasons, and the fruit is a significant element in local markets. Brazil is at the head of production for the world market, followed by Ecuador, Panama, Venezuela, and Honduras. In the East bananas are grown in quantity in the Philippines, Malaysia, India, and Indonesia.

Bananas and its botanical relatives are used locally in numerous ways. The pithy pseudostems ("trunks") of *Musa textilis* yeld a fiber (Manila hemp) that may be spun into cordage. The flower buds are cooked and eaten as a vegetable; and in Africa the large starchy seeds of certain species are used as food, and also strung as beads. In Malaysia, the broad leaves are sometimes a part of ceremonial ornamentation. Wax can be obtained from several species by boiling the leaves in water; the wax will separate and float to the surface. An alcoholic beverage is also produced by fermentation of the fruit. In the West Indies a cider is commonly made, the origin of which is believed to be African. Banana pulp, in combination with certain chemicals, may be used for dying fabrics, such as silk and cotton, but the quality of the dyes is poor. A low-quality paper is sometimes made from the stems. The leaves are used as roofing, wrapping for food and other items. In Thailand they are dried, cut, and employed as cigarette papers. Other occasional uses are as bandages, livestock food (especially for pigs), and as a source of potash. Bananas have a mystical connotation in some areas. In Malaysia they are planted in the corners of fields, to protect the harvest. In Singapore women, after giving birth, sometimes bathe in water in which banana leaves have been boiled, as part of the ritual pertinent to insuring the child's future.

Figs

Figs (*Ficus carica*), an ancient and important fruit of the Mediterranean, were until quite recently a staple of peasant cultures in Greece and the Balkan region. Tomb drawings more than 4000 years old depict the growing of figs in Egypt. Figs were brought to the Americas by the Spanish, and were often planted on the grounds of missions and other settlements. The leading cultivated variety in California, Mission figs, were introduced at the San Diego Mission in 1769.

Edible figs bear two crops of fruit each year, the breba or first crop on old wood and the second or main crop on current shoot growth. The wild, inedible caprifig produces a third overwinter crop. The fig fruit is a syconium consisting of a swollen hollow fleshy stem tip bearing on the inside many small individual flowers. Edible fig fruits contain only long styled female flowers and no pollen. Caprifigs have short styled female flowers and also male, pollen producing flowers. Edible figs such as the varieties Adriatic or Kadota are of the common type in which both fruit crops set without pollination. These fruits are without viable seed though the seed coat may become hard. Another edible fig, the Smyrna type, such as the variety Calimyrna, requires pollination for both the breba and main crop fruit with pollen from the caprifig. This pollen is transferred by a small female chalcid wasp, *Blastophaga psenes,* which can oviposit in the short styled caprifig flowers but not in the long styles of the edible fig. All edible figs and the caprifig can be pollinated, but only the Smyrna fig requires caprification which results in development of viable seed. Smyrna fig trees grew well in California but did not produce fruit until the caprifig and the Blastophaga wasp were introduced about 1899.

There is a very specific symbiotic relationship between the fig wasps, the caprifig, and Smyrna fig. The first generation of syconia in caprifig, usually formed in February, contains staminate flowers with pollen near the eye. Farther back are the pistillate flowers with short styles some of which contain insect larvae, the gall flowers. A female wasp enters the eye of the syconium and oviposits an egg each in several ovules which become the gall flowers. There the wasp larvae hatch and mature into adult wasps. Upon maturity the wingless males chew out of their chambers and proceed to locate gall flowers containing females where the male pierces the ovary wall, fertilizing the female within. The male wasps die within the syconia.

When mating has been accomplished, the female wasps having wings leave the syconia. In doing so they brush against the pollen flowers near the eye. The females with pollen now seek out the young fruits of second crop caprifig, or if in the vicinity of young Smyrna fig will enter the new syconia and attempt to lay eggs. Thus eggs are deposited in caprifig fruit with short styles but not in Smyrna fruit with long styles, but pollination of the Smyrna fruit is accomplished. Later in the fall the Blastophaga repeats the process as eggs are laid in the overwinter, third crop caprifig. Again in the spring the event is repeated.

The annual world production of figs is about one and a half million tons, and comes predominantly from Portugal, Italy, Greece, and Turkey. In the United States, California's yield is about 70,000 tons, most of which is grown in the San Joaquin Valley. Israel has been emphasizing fig production, as the trees are well-suited to that climate. The fruits are eaten fresh or dried. Figs are raised primarily for human consumption, but they are also used as stock feed in the Old World.

Breadfruit (*Artocarpus communis*) is an important tropical staple. Brought from Tahiti to the West Indies by Captain Bligh some time after his ill-fated voyage on the *Bounty*, the fruit is an inexpensive yet productive food source for plantations, and thrived in the American tropics. The trees are extremely temperature sensitive, and do not fare well even in Miami, Florida, where temperatures occasionally drop briefly to freezing.

The trees are attractive as well as productive, growing to a height of 60 feet, with spreading branches. The leaves are broad, dark-green, and leathery. The large fruit weigh up to 20 pounds; one to three fruits grow at the tip of a branch. There are two distinct kinds of breadfruit: the cultigen, which has no seeds, and a wild variety with large, chestnut-shaped seeds. In Micronesia, Melanesia, parts of Indonesia, and Polynesia, the fruit is an important part of the diet. It is baked, fried, or dried and reconstituted later with water. One of the starchiest of fruits (26% or more carbohydrate), its consistency is somewhat breadlike. The trees supply a few additional products. Cloth is sometimes woven from the fibrous inner bark; the wood is commonly an element in construction; and the milky sap is used as glue and caulking. The wild variety's seeds are a nutritious food, having a high protein content; usually they are roasted.

Other Widely Cultivated Fruits

Cherries *(Prunus spp.).* These fruits are today widespread in the northern hemisphere, but the stock giving rise to most cultivated varieties originated in Asia Minor at least several thousand years ago. Around 100 AD Pliny mentioned 10 types of cherries in Italy; and the trees are known to have been frequently grafted by Roman horticulturists. Introduced into the Americas during Spanish and English colonization, commercial cherry production in the United States is currently centered in Michigan, Washington, Oregon, and California.

Peaches *(Prunus persica).* This fruit is of Chinese origin. The names of the two major cultigens, "clingstones" and "freestones," refer to whether or not the flesh adheres to the seed. Peaches were brought to the Americas by the Spaniards, and were a popular fruit in missions. Trees in Canyon de Chelly, a Navajo center in northeastern Arizona, and some trees along the Rio Grande pueblos of New Mexico, date from such mission plantings.

Apricots *(Prunus armeniaca).* This fruit is mentioned in Chinese writings dating from 4000 BC. From China, the fruit was brought to southwestern Asia, and had reached Rome by 1900 BC. The Spanish took peaches with them to their colonies, and the English established the trees in Virginia. The name *armeniaca* refers to Armenia, which was probably the starting point for their dispersal into Europe.

Pears *(Pyrus spp.).*   A fruit of western Asian origin, 36 varieties were already being cultivated in Rome 1900 years ago. The cultivation of pears is problematic, as some trees will bear for more than 300 years. The hard and durable wood is used in the manufacture of musical instruments, and in some furniture construction. Pears (and also cherries, peaches, and apricots) are ingredients for delicious wines and brandies.

Melons *(Cucumis spp.).*   We are familiar with such common names as cantaloupes, watermelons, crenshaws, persians, honeydews, and cucumbers. These melons are probably of African or eastern Asian origin, although origins are uncertain. Cantaloupes were grown in Egypt 4400 years ago. Watermelons (*Citrullus sp.*) have long been widely distributed in Africa, and were brought to America during the time of slavery. Most melons are eaten raw, but Oriental cookery does make use of them, especially cucumbers and winter melons.

Most varieties prefer hot weather and a dry climate for ripening. Growing areas in the United States are primarily the southeast for watermelons, and southern California and Arizona for cantaloupes and others. The fruits, characterized by a thick rind, are specialized berries which grow on trailing vines.

Roots

Edible plant parts growing underground are usually referred to as "root crops," although by botanical definition they may, in fact, be modified stems or leaves condensed as a bulb, as well as true root. Roots are not as easily harvested as grains and cereals, because they must be unearthed. Nor are they as nutritious as grains, legumes, and even many fruits, especially in protein content. Yet several of the so-called root crops, such as Irish, or white potatoes, and cassava are among the most important human dietary staples.

One factor in their importance is productivity. Tens of tons may be yielded from a single acre, without the expenditure of continuous care. Although not as easily preserved as grains and some other foods, they are, on the whole, less perishable than a majority of other fruits and vegetables. Some are a good source of energy, which is available as carbohydrates. Thus, a diet based on potatoes or other starchy crops will lead to malnutrition unless supplemented with foods containing a higher proportion of protein.

Irish or White Potatoes   One of the more important human foods is the Irish, or white potato (*Solanum tuberosum*). Indigenous to South America, it was being cultivated in the highlands from Chile to Colombia by the time Spanish soldiers began trekking through the Americas. The first known European literary reference to potatoes appeared in 1552, in Pedro de Leon's *Cronica del Peru*, and the first published illustration was printed in Gerard's *Herbal* in 1633. Plants were taken to Europe prior to 1570, but acceptance was slow. Reintroduction to the New World, into North America, was made in 1621, probably by way of Bermuda. It was not until the 1700's, however, that potatoes began to spread as a crop. Some European governments, recognizing that the potato might provide a significant food source, required its planting; royal edicts of that nature were declared in Germany in 1744, and in Sweden in 1764. White potatoes, particularly, became a staple in Ireland, to the point that the *per capita* consumption among the peasantry was 10 to 12 pounds per day. So when the potato blight, a fungal disease caused by *Phytophthora infestans*, resulted in the failure of the Irish crop for two consecutive years in the 1840's, famine and migration to America followed.

The Irish potato is an upright, branching, and somewhat spreading annual that grows to a height of 2 or 3 feet. The roots are fine and fibrous, with numerous rhizomes that are swollen at the tip to form the edible tubers. They are adaptable to many soil types and climates, and are cultivated throughout the world except in low, tropical areas. The plants will thrive at altitudes of more than 10,000 feet, and are a food resource in regions where perennial cold retards or prohibits growing many other crops. Ideally, the environment should be cool and moist, with a mean annual temperature of 40° to 50°F. Potatoes grow best in well-drained soil.

Potatoes are generally propagated vegetatively, by tubers, or parts of tubers; the latter are known as "seed potatoes." Today, there are more than 500 varieties obtained through hybridization or by mutations. Since 1850, investigators have been experimenting to improve potatoes. A germ plasm reservoir is maintained at the USDA Experimental Station at Sturgeon Bay, Wisconsin.

A potato tuber is predominantly water, from 17% to 34% carbohydrate, a small quantity of protein, a trace of fat, and some vitamin C and minerals. The varieties usually cultivated in the northern hemisphere are very low in protein, containing only 1% to 3%, but some South American cultigens have as much as 7%. Unfortunately, much of a potato's food value is in its skin, which is often removed for processing or peeled when it is eaten fresh. Potatoes, which are related to nightshade, develop a toxic alkaloid, *solanine*, especially when the tubers have been exposed to light during growth.

The tubers keep fairly well, although they lose their crispness after several weeks. Frost is not beneficial to storage; but in the high Andes, where frosts are frequent, the Indians have used it to their advantage. Potatoes are left out to freeze at night, and then stamped on while thawing. After a few days of this treatment, most of the water has been removed, and the desiccated tuber, called a *chuño*, may be kept for long periods.

The annual world potato crop is currently more than 300 million tons, 90% of which is grown in Europe and the USSR. The latter leads in production, followed by Poland and Germany. The total European and USSR yield is about 230 million tons, followed by 15 million tons grown in the United States and 7 or 8 million in South America. Most of these crops are used as human food. In the U. S., more potatoes go into the manufacture of processed foods, such as potato chips and instant mashed potatoes, than are consumed fresh. Potatoes are also fed to livestock, especially in Europe. Starch is produced from them, and they are a source of alcohol, both for industry and for drinking.

Sweet Potatoes

Another root crop indigenous to South America is the sweet potato (*Ipomoea batatas*). At the time of Columbus, the sweet potato was widely distributed in South America and on some Pacific islands as well, having become a staple food of the Maoris in New Zealand.

A crop that is surely native to South America existing on distant Pacific islands poses a puzzle as to how such early distribution was accomplished. It is possible that seeds may have been transported on logs or by birds, but it seems more probable that human beings were the medium. The Polynesians, who were expert seamen, may have landed on the coast of South America and returned with sweet potatoes among their stores. It is also not implausible that pre-Columbian Peruvians, who also constructed sea-going boats, may have been blown off-course and been carried to a Pacific island, a theory supported by the direction of prevailing winds. In that case, a return trip would not have been necessary. Or, both possibilities may have occurred.

The origin of domesticated sweet potatoes is not clear. A

*A western Nigerian mother
crushes cassava. (FAO photo)*

wild ancestor was probably a plant native either to Mexico or Peru. The domestic species is hexaploid. A Japanese botanist, Ichizo Nishiyama, has discovered a wild hexaploid species (*Ipomoea trifida*) in Mexico that may be an ancestor of the cultivated one. On the other hand, the only archeological evidence of sweet potatoes was discovered in Peru.

Sweet potatoes grow much better in humid regions than do Irish potatoes. Like white potatoes, they do best in a well-drained soil. The root provides about 50% more calories per unit of weight than does an Irish potato. However, its popularity and commercial importance have remained minimal. Three to 6% of the 26% carbohydrate content is sugar, but there is even less protein than in white potatoes: 1% to 2%. Sweet potatoes are exceptionally rich in vitamins, calcium, and iron, so their nutritional value is at least comparable to that of most fruits and vegetables.

In more temperate regions, planting is usually accomplished by using shoots taken from "seed" roots retained from the previous year's crop. In tropical areas the vines grow more or less continuously year round, and propagation is by stem cuttings. Sweet potatoes are customarily planted in raised hills, sometimes between rows of a grain or some other crop. The yield may be as much as 20 tons per acre under optimal growing conditions, but because most of the processes involved in cultivation must be done by hand, the cost is high. Major world growing areas correspond to those places where field labor is inexpensive.

The annual world yield of sweet potatoes is not known precisely, because yields are not distinguished from the similar but unrelated yams (*Dioscorea spp.*). Together, more than 100 million tons are grown yearly. Japan leads in production, followed by other Asiatic countries. Sweet potatoes are a common table vegetable in the Orient but, to a lesser degree, in the United States. Besides being served fresh, they are canned, desiccated, manufactured into flour, and are also a source for starch, glucose, syrup, and alcohol. The roots serve as livestock feed, and the bushy green tops as fodder. Sweet potatoes spoil easily, causing a third or so of the crop to be lost every year.

Manioc

One of the most important tropical food crops is a plant relatively unfamiliar to us. *Manihot esculenta*, known variously as manioc, cassava, and yucca, belongs to the same family (*Euphorbiaceae*) as poinsettias and the Pará rubber tree, the best source of natural rubber. The plant may grow rather tall, to a height of 15 feet or more, but its edible part is a tuberous root reminiscent of sweet potatoes, although considerably larger. A single root may be a yard long and weigh several pounds.

Manioc consists of several vague varieties, divided between two groupings, sweet and bitter. The former may be eaten directly, the latter contains concentrations of toxic cyanogenic glucosides in sufficient quantity to be dangerous if ingested without preparation. The bitter manioc must be soaked or boiled to extract the poison. How did human beings first learn that an otherwise poisonous plant could be rendered edible by certain techniques of preparation? So far the question is unanswered, but bitter cassava was being prepared and eaten by Indians throughout South America when Europeans first arrived.

The roots are eaten raw or cooked, and sweet varieties are often boiled. A flour-like meal, *farinha,* is commonly made in South America by peeling, washing, and grating the roots, after which the material is placed in a tube of woven fibers—a *matapé*—and the liquid in the grated root squeezed out. Then the meal is dried, sifted, and pressed into thin, heavy cakes known as *cassava bread,* which has a reasonable food value. In this form it is a mealtime complement throughout eastern South America. Boiled, and somewhat tasteless, manioc is frequently eaten in Paraguay, parts of Brazil, the Andean countries, Cuba, tropical Africa, and the Far East. The juices squeezed out when making *farinha* are saved and fermented to be served as a beverage, or used in various sauces and the "pepper pot" dish local to the West Indies and Guyana.

In North America, manioc is known as tapioca, the small white pellets that are a basis for puddings. Tapioca starch is also used as a raw material to make mucilage. The starch is also hydrolyzed into simple sugars, syrups, and other products, and the residue is used as animal fodder.

There is no known wild variety of M. *esculenta.* The geographical centers of distribution are an area consisting of western and southern Mexico, parts of Guatemala, and northeastern Brazil. Archeological evidence suggests that manioc was grown in Peru 4000 years ago, and at least 2000 years ago in Mexico. By the time of the European conquest, its range of cultivation in the New World was as broad as it is today. During the last half of the 16th century, Portuguese sailors carried manioc to Africa, where it was of negligible importance until the 19th century. Today Africa grows more than any other area. Manioc was taken to Malaysia in the 1700's, and currently is cultivated in all tropical regions.

Manioc is primarily a lowland, tropical crop but will grow at an elevation of 5000 feet at the equator. The plant cannot tolerate cold and frost, but can tolerate rainfall varying from

20 to 200 cm per year. *Except during planting*, manioc is able to withstand prolonged periods of drought, and is a significant crop plant in regions of low or uncertain precipitation. Sandy or loamy soils of at least moderate fertility are preferred, although the root will grow in almost any soil that is not waterlogged, too shallow, or too stony. As it will thrive fairly well on land that has been exhausted by other crops, manioc is usually the last crop planted in successions of shifting cultivation. Estimates of world production are not easily determined, as most manioc is consumed locally in rural areas. Estimates place the annual crop at 75 million tons.

Other Root Crops **Beets** *(Beta vulgaris).* Other root crops include beets, which are separated into those grown for sugar, and the more delicate table beets. The latter are usually consumed when only a few weeks old, while the roots are still sweet and tender. Cultivated beets are presumed to be a derivation of a wild European relative *(Beta maritima)*. Chard, a type of beet, is eaten as a "green," just as were cultivated beets until the 16th century. Economically, sugar beets are of far more importance (see Chapter 10).

Carrots *(Daucus carota).* These were originally a weed of the family *Umbelliferae*. The cultivated variety was selected long ago in the Near East for its parsley-like top. The development of the root as a vegetable has occurred only within the past few centuries.

Taro *(Colocasia esculenta).* This root is an intrinsic staple in Polynesia and southeastern Asia and has numerous cultivated varieties. The plant grows from a fleshy corm which can be baked, boiled, eaten whole or mashed, or ground into a meal. The Hawaiian dish *poi* is taro that has been crushed and fermented. Adaptable to marshy, tropical soils, the large "elephant ear" leaves may be seen growing in fields in warmer parts of Florida, where the cultivation of taro has been increasing. A similar genus, *Xanthosoma*, the most commonly eaten species of which is *X. sagittifolium*, is native to South America. Its current distribution is pantropical.

Yam *(Dioscorea spp.).* This is the third most important tropical root crop, following sweet potatoes (with which they are often confused), and manioc. The tubers may grow to enormous sizes, weighing as much as 100 pounds. They are an expensive crop in terms of labor, because one must dig so deeply for the tubers. The vines of wild yams are found throughout the tropics. There are 600 to 800 species. In addi-

Above *Vegetative growth of the taro plant. The underground corms are used as a starchy food. (USDA photo)*

Left *Giant radishes are raised for cattle feed on government farms in Sri Lanka. (FAO photo)*

tion to food value, the plant is a source of the steroidal sapogenins used in the development of cortisone. In West Africa, crushed yams are a basis of the *fufu* dish. They may also be fried, baked, and boiled, or used as soup stock.

Horseradish *(Rorippa armoracia)*.   This member of the mustard family grows a white, carrot-shaped root from which a pungent glucoside, used primarily in flavoring, is obtained.

Jerusalem artichoke *(Helianthus tuberosus)*.   This is a relative of the sunflower and has a long tuber-like rhizome which is eaten boiled, pickled, or raw.

Allium spp.   Onions, garlic, leeks, shallots, and chives are important food plants in almost all countries. Onions are especially useful as a flavoring element, or boiled. Anatomically, they are true bulbs, composed of condensed food-storage leaves rich in sugar and sulfides, which give them the customary acrid quality. The major onion growing area in the United States is the muckland of the Great Lakes. Onions are also a major crop in central and southern Europe.

Parsnip *(Pastinaca sativa)*.   This is a relative of carrot, with long whitish roots that sweeten after exposure to cold.

Radishes *(Raphanus sativus)*.   One of the easier and quicker growing garden vegetables, radishes are native to China, where they have been cultivated for thousands of years. Of lesser importance in the United States and Europe, they are available in most produce markets.

Rutabaga *(Brassica campestris napobrassica)*.   Probably originating as a cross between cabbage and turnip, the yellow-fleshed root is reportedly more nutritious than that of turnips. Consumption is confined largely to Europe and North America.

Turnip *(Brassica rapa)*.   This is a biennial of the cabbage group, and has long been cultivated in central Asia and Europe. Turnips are frequently used as a livestock feed.

Suggested Readings

Heiser, Charles B. *Seed to Civilization*. San Francisco: W. H. Freeman and Co., 1973.

Herklots, G. A. C. *Vegetables in Southeast Asia*. London: George Allen & Unwin, Ltd., 1972.

Hill, Abert F. *Economic Botany*. New York: McGraw-Hill Book Company, Inc., 1952.

Janick, Jules. *Horticultural Science*. San Francisco: W. H. Freeman and Co., 1963.

Mortensen, E., and Bullard, E. T. *Handbook of Tropical and Sub-Tropical Horticulture*. Washington, D.C.: U. S. Dept. of State, AID, 1964.

Purseglove, J. W. *Tropical Crops: Dicotyledons I*. New York: John Wiley & Sons, Inc., 1968.

Salaman, R. N. *The History and Social Influence of the Potato*. Cambridge: Cambridge University Press, 1949.

Scagel, R. F.; Bandoni, R. J.; Rouse, G. E.; Schofield, W. B.; Stein, J. R.; and Taylor, T. M. C. *An Evolutionary Survey of the Plant Kingdom*. Belmont, California: Wadsworth Publishing Company, Inc., 1965.

Thompson, H. C., and Kelley, W. C. *Vegetable Crops*. New York: McGraw-Hill, 1957.

CHAPTER

9

THE INCREASINGLY IMPORTANT LEGUMES AND NUTS

Although cereals may be given credit for the rise of civilization, the legumes, of the botanical family *Leguminosae*, run a close second. They are higher in protein content than any other plant product, and therefore more nearly approximate animal flesh in terms of their food value. Also, the amino acid content of the legume protein complements the amino acid content of the cereals' protein. Thus a diet with both legumes and cereals has a more "complete" protein content than a diet of either legumes or grains alone. There is a nutritionally sound basis for the staple ethnic diets of many areas; in each case a grass and a legume are involved: corn and beans or rice and beans in Latin America and the Caribbean; rice and soy beans in the Orient; wheat and peas in Scandinavia.

Legumes

Legumes appear in the archeological record as domesticates almost as early as grains. Human beings were collecting the seeds of wild legumes while growing the first cereals as crops. Cultivated lentils are present in archeological deposits in the Old World dated at 5000 BC, and in the New World beans have been dated at 5000 BC in Mexico and 1000 BC in Peru. Soybeans have long been one of the most important crops in the Orient—providing the basis for a broad variety of foods.

It is questionable whether any of the early Old or New World civilizations would have arisen had there been only grasses, instead of the dietary union of grains and legumes. Adequate protein, not only in quantity but in the presence of all amino acids necessary for human growth, is essential if people are to realize their physical and mental potential. *Kwashiorkor*, a debilitating disease caused by protein deficiency, is common in parts of the world where high-starch diets are not sufficiently complemented with legumes or meats. If the Maya had only corn, it is doubtful that they would have developed sophisticated civilizations. Quantitative analyses have shown that *zein*, the principal protein of corn, and a and b globulins of black beans, the other staple in the Mayan diet, are complementary. Also, beans are high in lysine, which is the amino acid lacking in *zein*.

Soybeans and peanuts, for example, are nearly as rich in protein as whole dried milk or powdered eggs. Animal flesh, ranging from 70% to 90% protein dry weight, is far more valuable. But meat is expensive and not available in quantity to much of the world's population. Legumes are a most important substitute.

As a group legumes are high in oil as well as protein. Soybeans contain 20% oil; peanuts 50%. In countries such as the United States, where protein scarcity has not yet become a significant problem, legume crops are cultivated mainly for the oil, with protein as a by-product. In the U. S., 90% of soybean oil goes into the preparation of processed foods, such as margarine. The remainder is used mostly for industrial purposes.

Legumes are also used as livestock feed, in amounts nearly equal to grasses. Legume leaves are valuable forage because of their high protein content. Alfalfa, probably the oldest cultivated forage plant, is high in protein as well as yield. Seeds more than 6000 years old have been uncovered at archeological sites in Persia, proof that alfalfa has long been a crop. Today, alfalfa is grown almost exclusively as a livestock food, due to a flavor generally unappealing to humans.

Nitrogen Fixation Legumes are also important in maintaining and improving soil fertility. Although this quality has long been recognized, it was only toward the end of the 19th century that the reason was determined: legumes add nitrogen to soil. Most legumes have nodules on the roots which contain bacteria that are able to "fix" atmospheric nitrogen—thus making nitrogen available to the host plant. In exchange, the bacteria are supplied with carbohydrates. Later, when the nodules disintegrate, nitrogen is returned to the soil. In agricultural areas where land must be continuously cultivated, growers alternate legume crops with others to revitalize soil. The root nodule bacteria (Rhizobium spp.) lives freely in the soil unless legumes are available. Then the bacteria is attracted to the legume roots, entering through the root hairs and into the cortex. The presence of the bacteria causes cells to divide, producing the characteristic nodules. The growth and efficiency of these nodules is determined by the carbon-to-nitrogen ratio of the plant and by the amount of phosphate, calcium, magnesium, molybdenum, and boron in the soil. The number of nodules on the roots has no direct correlation with the nitrogen-producing ability. If for some reason, the nodules are ineffective, the bacteria will derive nitrogen from the host plant. In this situation, the legume plants must then obtain nitrogen from the soil, and the soil will be exhausted even faster than by cultivation of a grass crop. Most research on nitrogen fixation applies to crops grown in temperate regions. Relatively little is known about the value of leguminous plants in sustaining soil quality in tropical areas.

Pulse Domestication in the Near East and Europe *Pulses* are the edible seeds of peas, beans, lentils, and related plants having pods. In the Old World Mediterranean agricultural belt, various pulses—

Above *White Dutch Clover is an important legume fodder crop. Nitrogen is fixed by the bacteria in the root nodules. (USDA photo)*

Left *Cultivation of peanuts. (FAO photo)*

peas, lentils, broad beans, and chick peas—are universally in company with wheat and barley. Whereas the domestication of grasses has been intensively studied, there is minimal archeological data on pulses. But during the past few years, a significant quantity of carbonized remains of legumes have been discovered at Neolithic and Bronze Age sites in the Near East and in Europe. These archeological finds have been further clarified by botanical and genetic studies of wild legumes related to domesticated ones.

Peas *(Pisum sativum).*   Peas first appeared in the archeological record of Neolithic farming villages in the Near East. These sites are dated at 9000 to 8000 BC (see Zohary and Hopf 1973). Pea seeds were obtained from Jarmo in northern Iraq, Cayonu in southeastern Turkey, and pre-ceramic levels of Jericho. Later Neolithic sites in the Near East, dating from 6000 BC, yielded far more pea remains, among the wheat and barley. Peas are also abundant in European Neolithic sites.

Most pea remains at archeological sites are in the form of carbonized seeds. Unfortunately, such remains are not as helpful to investigators as are the remains of grains. Archeologists cannot unravel the history of the legume's domestication from the carbonized seeds. In the Near East, a criterion for cultivation of cereals is the presence of a nonbrittle rachis (see Chapter 6), but there is no clearly diagnostic character for determining whether or not a pea seed is from a wild or cultivated plant. Cultivated peas are usually larger than wild ones, and with domestication the *hilum** length increases. However, these changes are subtle, occurring over long periods. Accordingly, there may be a great deal of overlapping between wild and domestic varieties. Possibly the surest means for separating wild pea remains from domesticated varieties is the nature of the seed coat. In wild peas the coat is usually rough or granular. In contrast, the coat is smooth in domesticated varieties. Smooth peas dating back almost as far as domestic barley and wheat have been excavated—suggesting that the cultivation of pulses began at approximately the same time as grasses.

Lentils *(Lens culinaris).*   These are another important pulse crop in the Old World. Lentils were apparently domesticated at least as early as barley and wheat, and have been in use in the Near East even before there were farming villages. In Mureybit (8000–7500 BC), a primitive settlement in northern Syria, lentils appear to have been gathered from wild plants along with wild wheat and barley. In this pre-agricultural

*hilum: *A scar on a seed marking the point of attachment of the ovule.*

community, the earliest cultivation of lentils, wheat, and barley may have begun at the same time. Later in the archeological records, lentils were common from Near Eastern farming settlements (about 7000 BC). A few small lentil seeds were found at Jarmo; and some have been found in Anatolia, Iran, and other areas of the East. Later Neolithic settlements yielded greater quantities.

It is perhaps even more difficult to distinguish the seeds of wild from domesticated lentils than to recognize wild from cultivated peas; the seeds of wild lentils are structurally very similar to those of domesticates. The only major difference is in size, cultivated varieties being larger.

The cultivation of wild *Lens* is probably as old as agriculture itself. This fact can be inferred from the plant's ecology. Under natural conditions, lentils do not grow in stands of any size. The collection by early human beings of quantities of seed from the small, sparsely populated wild plants would have required far more energy than the yield warranted. Also, wild lentils are today unknown in the vicinity of the sites where remains have been uncovered, such as at Jericho. From these facts, one may conclude that the inhabitants of any site where many lentil remains are present were probably cultivating them. By the time Neolithic agriculture had become integral to human cultures in the Near East and Europe, the lentil seems to have been a part of it.

During the Neolithic and Bronze Ages of Europe and the Near East, several other pulses were domesticated. Broad beans (*Vicia faba*), bitter vetches (*Vicia ervillia*), and chickpeas (*Cicer arietinum*) are present at archeological sites. Little is known about the domestication of these. The bitter vetches probably were brought into cultivation in a manner similar to that for peas and lentils. The data on chickpeas and broad beans is inconclusive.

The Domestication of Beans in the Americas   There is documentation that beans as well as corn were well established as crops before the Spanish conquest of Mexico. The Viceroy Antonio de Mendoza, who served Spain in the New World from 1535 to 1550, compiled a manuscript based on older documents. Entitled *Codex Mendoza*, the manuscript lists items of tribute that were being paid to the last Aztec emperor, Moctezuma Xocoyotzin II, at the time of the Spaniards' arrival. These mandatory tributes included 280,000 bushels of maize per year, and 230,00 bushels of beans (Wolfe, 1959). Another estimate (Peterson, 1962) indicates that the emperor was receiving about 5000 tons of beans annually.

In the New World today, four indigenous species of beans

(*Phaseolus*) stand out as food crops. The common bean (*P. vulgaris*) is a polymorphic, poorly understood species, of which there are hundreds of cultigens—including red kidney, pinto, and navy beans. It is the most commonly cultivated bean, grown from sea level to heights over 2000 meters. Sieva beans (*P. lunatus*), are grown throughout the Americas. This species includes the broad, flat lima beans common in the United States. Runner beans (*P. coccineus*) have been developed as an indigenous crop most extensively in the highlands of Chiapas and Guatemala. In the United States, they are often grown as an ornamental.

Domesticated beans differ in several ways from wild species. Our knowledge of the processes of change from the wild to domesticated varieties is incomplete. As with peas and lentils, the seeds of cultivated beans are almost invariably larger than those of wild types. The oldest archeological bean materials discovered so far do not exhibit size differences from those varieties presently under domestication. This suggests that the transition in size occurred at some time prior to the earliest excavations, and perhaps even before the initiation of agriculture. An association of early corn and beans at the Mexican site of Tehuacan, more than 6000 years ago, indicates that beans were probably interplanted with corn. Large seeds may have given the bean plants an advantage in competing with corn.

The inability of bean seeds to take in water, a condition known as "hard seed," is an agronomically significant characteristic. When many seeds in a planting have this trait, crops will germinate unevenly. Those which absorb water easily will begin to grow ahead of those which do not. Dietarily, hard seeds are not advantageous. Beans are usually soaked in advance prior to cooking. When hard seeds are mixed in with more absorbent seeds, some are pre-softened by soaking and others not. Boiling beans without a preliminary soaking, which is commonly done in Mexico, affects their food value. Heating the beans dry at cooking temperatures will cause the formation of indigestible protein-carbohydrate complexes. Domesticated beans appear to have more permeable seeds than wild ones. Possibly, this is the result of early selection based on the slowness with which impermeable seeds germinate.

Most cultivated legumes have developed indehiscent seed pods which remain closed on maturity. Such pods are no longer capable of automatically dispersing seeds. Wild beans and other legume pods split along both sutures when mature —forcibly expelling the seeds. In contrast, some cultigen pods will split, but with little of the force characterizing the wild legumes. The absence of forceful seed expulsion is due to a reduction in cultivated varieties of the pod's parchment layer.

The earliest remains of unequivocally domesticated beans are from the Coxcatlan cave of the Tehuacan Valley in Mexico. A well-preserved, uncharred pod valve was taken from material having a radiocarbon date of about 5000 BC. The lack of a developed parchment layer is proof that the pod was from a domesticate. Other specimens of early beans have been discovered in Ocampo, Tamaulipas. In addition to structural evidence of domestication, the presence of the pods in caves that were human habitations is further evidence that these legumes were a cultivated variety. The pods of wild species would have split and evacuated the seeds.

Soybeans

The soybean (*Glycine max*) has been a principal food plant in the Orient for millennia; only in recent years has it become of much interest in the West. Historical and geographical evidence supports eastern China as the place of origin. Soybeans probably began to emerge as a cultivated plant over 1000 BC. The first mention in Chinese literature dates back to that period, but soybeans may have been cultivated much earlier.

Soybeans have become increasingly popular during the past few decades, because of the need for more high-protein, high-fat food to meet the demand of an exploding human population. In 1930, the United States produced 14 million bushels of soybeans, from a million acres of farmland. In 1973, 56 million acres were given to soybean cultivation, which yielded one and a half billion bushels. Soybeans have rapidly assumed a place of importance among U. S. exports. Twenty-five years ago, the export value of soybeans was negligible. In 1973 the export value exceeded 3 billion dollars, 5 percent of the total U. S. income from exports. The indication is that the U. S. soybean crop will become increasingly important in the future.

The expanding importance of soybeans is mainly due to the legumes' oil content and exceedingly high protein content. Eighteen percent to 20% of the crop is raised for oil production, a lower percentage than for other oilseed crops. However, in the case of soybeans, net production may be higher because the yield per acre is so great. Furthermore, protein content is as high as 30% to 40%, an exceptionally large quantity for any plant crop.

The soybean (*Glycine max*) is believed to have been derived from *G. usurriensis*, a twining, slender vine which grows wild throughout Asia. Although soybeans are sub-tropical plants, cultivation now extends to a latitude of more than 50°N. The plants' climatic requirements are similar to those of maize, and the most extensive cultivation in the United States is done in the "corn belt." Soybeans grow most prolifically in areas having a warm, damp summer; the plants cannot withstand the stress of excessive heat, or severe winters. A spectrum of soil types is suitable, although soybeans thrive in sandy or clay loams and alluvial soils of at least moderate fertility. The time needed for maturity is important for soybeans. In the United States, no less than nine "maturity groups" are recognized, each with a narrow range of limitations.

The first economic surge of soybeans in the United States occurred during World War II when butter and related products were scarce. Soybean oil became an increasingly larger constituent of margarine, causing cultivated acreage to rise to 10 million by 1943, 30 million in 1954, and 40 million in the late 1950's. When, in 1973, a total of 56 million acres of soybeans were under cultivation, the crop had assumed a major status. For a cultigen to move from a miniscule place to one of predominance in a large country's agricultural production within only four decades is a success story with few parallels.

Presently, the United States is the leading grower of soybeans, followed by China. On a worldwide scale the Chinese crop is not very significant, as most of it is consumed intranationally. Brazil exports the largest quantity of soybeans, second to the United States. The Brazilians concentrated on the development of soybean cultivation because that crop is amenable to the humid agricultural land. Other leading producers are the USSR, Japan, and Indonesia; although each of these produces only about a million tons per year, a small fraction of the United States' crop. Mexico, Colombia, and some other countries are making efforts to expand soybean cultivation, but as yet they are in the initial stages and have no effect on the world market. Hopefully soybeans will gain popularity in all tropical countries, for which the plant is well adapted. Widespread soybean cultivation would be a step toward solving the problem of protein deficient diets with which so much of the world must cope.

The increased consumption of soybean products in the United States is reflected in the popularity of margarine. In 1940, about seven units of butter were used for every one of margarine. Today, twice as much margarine is sold as is butter, and soybean oil accounts for three-fourths of U. S.

margarine production. However, margarine requires only a fifth of the soybean oil manufactured; the remainder is used in salad dressings, shortenings, a variety of nonfood products, and as export material. Soybeans are currently the largest single source of edible fat, constituting about 15% of the total world supply.

With the economic rise of soybeans, agronomists have paid more attention to them. New varieties, suitable for differing environments, have been developed; and researchers are experimenting to increase yields. The international community has benefited from U. S. research studies—especially Mexico and India, where the population suffers from one of the more severe malnutrition crises of any nation. Some American soybean varieties are producing higher yields when grown in India than the same varieties do grown in America. This is probably because of the compatible humid, tropical, and subtropical climates.

A variety of food products are already made from soybeans, and presumably their acceptance will increase. The beans are not popular as a main dish due to their distinctive and somewhat strong flavor. However, numerous synthetic meat foods are manufactured from soybeans. Protein isolated from soybeans has been available in the United States since 1960. Within ten years production of protein has reached an annual total of 20,000 metric tons. The protein is spun into fibers, which are then used to create artificial hamburger patties, steak, and other products meant to be acceptably similar to meat. It is probable that more refinements will further improve the taste of soybean synthetics—increasing popularity as a common food. Today, soybean meal is frequently added to hamburger, sausages, and other processed meats.

In the future soybeans may become a major food source, not only in developing countries but in the United States and other affluent nations. World food resources are of increasing global concern. All of us may have to accept changes in our customary diets. For example, bacon requires 10 times as much cropland as the soybeans needed to provide the same amount of protein. Beef production requires 15 to 20 times as much land. Soybeans and corn would provide a complete protein which, supplemented with occasional animal products, is nutritionally acceptable. Also, there would be less ecological damage that results from heavy use of nitrogen fertilizers; soybeans, being legumes, fix nitrogen from the air in root nodules. Clearly, energy consumption in the United States must be diminished, and meat production is an extravagant use of energy. Increased reliance on a soybean diet is one means of conserving energy through a change, not a reduction, in the quality of life.

Peanuts

Peanuts (*Arachis hypogaea*), also known as groundnuts, goobers, or pinders, are probably the second most valuable legume nutritionally. The high protein (26%) and oil (50%) contents make these legumes a valuable food source. Until recently the origin of peanuts was thought to have been in China. Now it has been discovered that South America, probably the Gran Chaco of Paraguay was the place of earliest cultivation. From South America, peanuts were carried to other parts of the world by Portuguese sailors, and finally were introduced into the state of Virginia by slaves transported from Africa. Archeological specimens more than 3000 years old have been found on the coast of Peru—indicating that peanuts are a human food of some antiquity. Two thousand year old samples have also been found in Mexico, suggesting that peanuts were important for trade or were included among food plants taken on migrations. By the end of the 16th century, peanuts had been carried by European explorers to China, Japan, Malaysia, India, Madagascar, and Africa.

The fruit is unusual, as it first appears on the end of a pointed, stalk-like structure (the *carpophore*); but as the seed-pod is formed, the peanut is forced underground by elongation of the pedicel, and there matures. The plant is a bushy or creeping annual. The creeping type spreads widely on the ground, and so is customarily planted far apart. The bushy one grows in a contained bunch. Most of the modern cultivated varieties exhibit characteristics intermediate between these two types. Peanuts grow well in loose, friable, well-drained sandy loams high in calcium and containing at least moderate quantities of organic matter. The plants do not tolerate frost; death usually results from exposure to cold. Peanuts require at least 40 inches of rainfall annually, with 20 inches during the growing season.

Annual production of peanuts in the United States is a million tons, grown mostly in Georgia, North Carolina, Texas, Alabama, and Virginia. Approximately half of this yield goes into peanut butter; 10% to 20% is used for oil. Total world production is 15 million tons or more, most of the crop being grown in India, Africa, and the Far East. China, Nigeria, and Senegal also grow respectable quantities. Most peanut oil is produced in Africa and India. It is used in margarines and shortenings, as salad or cooking oil, in soaps, and for industrial purposes. Cakes pressed in the extraction of oil are rich in protein and are a valuable livestock food. As with soy-

Above The recurved pericarp
of the cashew fruit contains
the nut. It is attached to a
fleshy receptacle which is
eaten as a fruit in areas where
the nuts are harvested. (USDA
photo)

Left Pistachio nuts ready for
harvest. (USDA photo)

beans, the use of peanuts will probably become broader. During the early part of the 20th century, George Washington Carver experimented effectively with hundreds of uses for peanut products. Today, peanuts are employed in the manufacture of plastics and numerous other goods. Future importance, however, is more apt to be related to the high-protein, high-energy quality as a food.

Nuts

"Nut" commonly refers to the seeds or fruits of any of a number of plants, such as peanuts (legumes) and coconuts (palms). Botanically, the term is defined as a hard-walled, indehiscent, one-seeded fruit. Nuts are not of great significance commercially, although they represent a concentrated source of food, usually rich in fat or oil, and with reasonable amounts of protein and carbohydrates. Nuts can be easily stored, and are not difficult to transport. Spoilage is a minimal consideration.

Generally nuts are not relevant to the problem of feeding humanity. Occasionally, in the past some have been food staples. In the American southwest, the seeds, or "nuts," of piñon pines were a highly important Indian food. These are still collected today from wild stands of pines. Wild trees remain the source of most black walnuts and Brazil nuts. Other nuts such as pecans, filberts, and almonds, have been cultivated; and yields and fruit quality have been improved through selection.

Nuts are used for various purposes other than as human food. They are sometimes toasted and ground as substitutes for coffee, and are a nutritious food for livestock. Probably the main use in affluent countries is as a "snack" item. Nuts are frequently served with before-dinner drinks and at parties. In the modern world, they constitute more of a luxury than a staple, and are a relatively expensive food.

Depending on the type, nuts are fairly high in fat, protein, and carbohydrate content. Those high in fat include Brazil nut, cashew, hazel nut, hickory, pecan, *pili* nut, macadamia, and walnut. High-protein nuts are almond, beechnut, and pistachio, with almonds having the greatest international commercial value. Large amounts of carbohydrates are present in acorns and chestnuts. Acorns are often given to hogs as fatteners and were eaten in the past by American Indians, who ground them, leached out the tannins and other bitter elements, pounded them into a meal, and ate them as porridge, mush, and in other ways. In fact, acorns are still a major food item for poorer people in some Mediterranean regions.

Since nuts can be stored and shipped as easily as most grains and cereals, why are nuts not a valuable commercial food on the international market? A principal reason is that, in most cases, mechanical harvesting cannot be effected. As a result, the cost is great. Also, the yield tends to be much less per acre than that of cultivated cereals and legumes; and, generally, the food value of nuts is more in the oil than in carbohydrate or protein contents. In general, nuts are nutritionally inferior to the cereals and legumes. However, many nut trees will thrive on land of minor agricultural potential. With food shortages increasing, the status of nuts as a food crop may increase in the future.

Suggested Readings

Gentry, Howard Scott. "Origin of the Common Bean, *Phaseolus vulgaris*." *Economic Botany* 23 (1969): 55–69.

Hymowitz, T. "On the Domestication of the Soybean." *Economic Botany* 24 (1970): 408–421.

Kaplan, Lawrence. "Archeology and Domestication In American *Phaseolus* (beans)." *Economic Botany* 19 (1965): 358–368.

Peterson, F. A. *Ancient Mexico*. Britain: Capricorn Books, (1962).

Pitkanen, A. L. *Tropical Fruits*. Lemon Grove, California: Renan Prevost, (1967).

Wolf, E. R. *Sons of the Shaking Earth*. Chicago: University of Chicago Press, (1959).

Woodroof, Jasper Guy. *Tree Nuts*, 2 Volumes. Westport, Connecticut: The Avi Publishing Company, 1967.

Zohary, Daniel, and Hopf, Maria. "Domestication of Pulses in the Old World." *Science* 182 (1973): 887–894.

CHAPTER

10

THE RECENT ROMANCES OF SUGAR AND LATEX

Sugar Cane   Sugar cane (*Saccharum officinarum*) is a grass which was probably first cultivated in New Guinea and Indonesia, and was distributed from those points to most of tropical Asia. Alexander the Great brought sugar cane back to Europe from one of his Asian expeditions. He was intrigued by a report that in Asia people could obtain honey without bees; thus he learned of sugar cane. Columbus supposedly transported sugar cane to the New World. Before the dispersion of sugar cane during the past five centuries, honey and some fruits were the only sweeteners known to most peoples. Sugar beet cultivation, today the second most important source of sugar, did not begin until the 19th century.

The annual world production of cut sugar cane is about one half billion tons, from which some 60 million tons of sucrose is extracted. Major centers of cultivation are Africa, the Caribbean, Brazil, Mexico, Hawaii, the Philippines, Indonesia, and Louisiana and southern Florida in the United States. Lesser quantities are grown in the Mediterranean and in a few other countries having an appropriate climate. The expansion of sugar cane cultivation in the United States began when the Cuban sugar embargo was imposed in the 1960's.

The plant is a rapid-growing, hardy grass that reaches a height of as much as 20 feet. Propagation is usually from cuttings, which root easily. Prior to mechanization, sugar cane was planted, tended, and harvested by hand. Today, in small holdings, it is still often cultivated this way. On large, modern sugar plantations, planting machines are used. Such machines accomplish the entire planting process—opening furrows and depositing and covering stem sections. The cane is usually harvested when it is a year and a half old. This was once done entirely by hand. Now huge cutting machines are used by large operations. However, in many areas of the world cutting is still done by machete. In southern Florida, the grueling, demanding work is done mostly by migrant laborers.

Sugar cane is a perennial, and will generate new stems, without human intervention, for years. However, replanting is often done after the third harvest, to insure the crop's vigor. The cane is cut into four-foot lengths and transported to a mill or factory, where the stems are shredded and crushed between rollers to extract the juice. The juice is then heated and mixed with lime to create a scum of impurities which can be skimmed off. The liquid is then concentrated by boiling, until it has reached a stage known as *massecuite*, a combination of syrup and sugar crystals. The crystals are separated centrifugally. The syrup becomes molasses; the latter, plain raw sugar. The cane residue, *bagasse*, is frequently used as a fuel in the mills, as a fertilizer, and as a component of fiberboard.

Raw sugar is brown in color and contains a respectable percentage of iron and protein. It is a staple in many tropical countries, but is almost unobtainable in the United States, where refined white sugar and synthetic "brown" sugars are consumed. The brown sugar is, in fact, caramelized white sugar.

Sugar is a major ingredient in candies, pastries, and a myriad of other Western foods. It is customarily available on dining tables, as a sweetener for tea or coffee. The molasses is also used in manufacturing candies, and is fermented into a rum. In tropical countries, molasses is also a source of alcohol and vinegar. Possibly sugar cane will become more important in supplying an energy source for the growing human population, but it will not answer other nutritional needs.

Sugar Beets   The second most productive source of sugar is the *sugar beet*, a variety of the common beet (*Beta vulgaris*) originally developed for its sugar in Germany. Two-fifths of the world's commercial sugar, about 23 million tons, is obtained from these. Sugar beets have been consciously improved to yield increasingly higher quantities of sugar. In 1774 the German scientist Marggraf first demonstrated the sugar beet's potential. From that time selection was made for those beets having a higher sugar content. Over the years, content was increased from 2–4% to 15–20%. During the early 1800's Napoleon favored expansion of sugar beet cultivation. This was successful largely due to Napoleon's edict severing trade with England—thereby cutting off sugar cane importation from the English colonies.

Sugar beets yield a net annual harvest of more than 160 million tons, most of which is grown in Europe, the U. S., and the USSR. As with sugar cane cultivation, the Cuban sugar embargo caused the expansion of sugar beet growing in the United States. Currently, the U. S. is second only to Russia in production, followed by Germany, France, Poland, and Czechoslovakia.

The beets will grow in almost any reasonably fertile soil, including those with saline content and those of semiarid regions—if irrigation is practiced. Plants are usually grown from seed, then are thinned out until they are 8 or 10 inches apart. Continual weeding and deep cultivation are necessary for crop success. Today, sugar beets are easily cultivated and harvested by machine. Sowing takes place in the spring, usually in April, and the plants are left to mature until October. The longer the beets remain in the ground, the greater is

Above Sugarbeets are a commercial source of sugar for refining and an important cattle fodder in the western United States. (USDA photo)

Right Mostly unmechanized, the cutting of sugar cane requires an enormous amount of manual labor. (FAO photo)

the sugar content. Upon being unearthed, the tops are cut immediately, to prevent the plant's utilization of the sugar.

Because of the soft, pulpy quality of the beets, the extraction of sugar involves a simpler process than that required by sugar cane. In earlier times, the beets were grated into a pulp and the juice was squeezed out into bags. Today, the beets are cut into strips, which are then placed in heated running water flowing through a series of tanks. In this manner, 97% of the sugar is removed. Impurities in the solution are precipitated out by carbonation, using lime to coagulate some of the nonsugars, and carbon dioxide to precipitate calcium carbonate, after which the juice is filtered. The procedure may be repeated several times, sulfur dioxide being added to adjust the alkalinity. The clear, filtered liquid is condensed, crystallized, and centrifuged, as in the cane sugar process. The pulp, either wet or dry, is used as feed for cattle and sheep; the filter cakes are valuable as manure; the discarded tops can be used as livestock feed or mulch; and the molasses is fed to stock or used in the production of industrial alcohol.

Maple Sugar

A third source of sugar, part of American tradition but of considerably less economic significance than the previous two, is the sugar, or hard maple, *Acer saccharum*. The collection of sugar from maple trees originated among North American Indians. The practice was not adopted by white settlers until colonization was well underway. The first known record of colonists using the technique was in 1673.

The Indian approach was uncomplicated. They roughly slashed the bark of the maple trees, and then collected the dripping sap in wooden or pottery vessels. The sap was concentrated by one of two methods: the particular boiling method of dropping hot rocks into it, since the Indians had no containers able to withstand extended heating; or by freezing, followed by a daily removal of the upper layer of ice that had formed overnight. White settlers began drilling holes instead of hacking slashes in the trees' trunks. They concentrated the sap by boiling it in large iron kettles.

Sugar maples are common in the forests of the northeastern United States. Sap may begin to flow as early as February, or as late as April, depending on when cool, crisp nights begin to precede warm, sun-filled days. The sugary sap is a result of the conversion of starches accumulated during the previous growing period into sugars during the winter, primarily in ray cells. The 2% to 6% sugar content of the sap is chiefly sucrose. The sap flow averages 34 days, although it may run for as many as 57 or as few as 9 days.

Above *Maple sap being collected in Massachusetts. The sap is taken to the sugar house where it is boiled down to form maple syrup and maple sugar. (USDA photo)*

Left *A rubber tree tapped for liquid rubber. (FAO photo)*

In contemporary maple sugar gathering, one to four holes per tree are drilled, at a depth of two inches. Metal spouts called *spiles* conduct the sap into pails, as rapidly as a drop per second. Formerly, people carried yokes with pails over their shoulders to collect the sap. Today the containers are assembled on sleds or power vehicles. Final processing takes place in a sugar house, where there is a shallow-pan evaporator 20 to 30 feet long. There is a wood fire at one end. There the sap is boiled down to a specific sugar concentration, and subsequently filtered for removal of accumulated limy materials. The end product is maple syrup, which weighs 11 pounds per gallon. Further boiling yields a thick paste that congeals and crystallizes, forming maple sugar.

The current annual production of several million gallons of syrup, almost all of which is obtained from wild trees, is likely to decline. Forests are decreasing in extent, and the cost of production is increasing. Already, many commercial "maple syrups" are heavily adulterated with cane sugar. Today, one of the larger consumers of maple products is the tobacco industry, which uses maple as a popular flavoring.

Rubber

Rubber trees (*Hevea brasiliensis*) were the source of 95% of the world's rubber until synthetic kinds were invented. Like sugar, rubber is a relative newcomer as the source of a widely used plant product, although the plant had long been used locally. Word of a strangely soft and resilient material used by the Aztecs was the first report Europeans had of rubber. Thereafter they learned of the Amazonian *caoutchouc*, from which the Indians made balls, watertight vessels, and even torches. In South America, by 1600 both the Indians and the European settlers used latex to waterproof clothing, houses, shoes, and other articles. However, rubber remained of no particular importance on the world market. Everyone believed that the uncoagulated latex had to be used *before* it hardened. Then, in 1823, the Englishman Charles MacIntosh discovered that solid rubber is soluble in naphtha, and began experiments on waterproofing with the material. In 1839, Charles Goodyear learned the process of vulcanizing rubber, by adding sulphur to it at appropriate temperatures. Vulcanization was the key to the industrial use of rubber, for which a myriad of applications were found.

When rubber first came into demand, it was collected only from wild trees. In fact when the automobile was invented, wild trees were still the sole source. This made the price rather steep. In 1910, natural rubber cost $3.60 per pound. Today, a pound of synthetic rubber costs about 30 cents.

In 1875, Sir Henry Wickham of the British Colonial Service

sent 7000 seeds of *Hevea* from the lower Tapajos of the Amazon to England. The young plants were germinated in Kew Gardens, and then transported to Ceylon. There, the plants were used to begin that region's existing rubber industry which, although slow in starting, eventually grew into thousands of plantations in Ceylon, Java, Sumatra, and Malaysia. In the early 1900's, the world's rubber capital was Manáos, a thriving, jungle-encircled city on the edge of the Amazon River. By 1914, the Malaysian plantations were productive, and soon overshadowed the South American market. For several decades, natural rubber, obtained mainly from the Pará rubber tree (*Hevea brasiliensis*), was an essential ingredient in the manufacture of tires and many other industrial products. Then in the early 1940's techniques were developed for fabricating synthetic rubber. Today, natural rubber comprises only about 30% of the total annual world rubber production. Of that 30%, 90% comes from Malaysia; Brazil and West Africa supply the remainder.

The indigenous habitat of Pará rubber trees is the low, hot, and humid forest region of the southern tributaries of the Amazon River. An estimated 300 million rubber trees grow in this area, where the climate is optimal. There are no temperature extremes, the usual range is 75° to 90°F, and the rainfall is 80 to 120 inches annually. Sometimes the trees live to be more than 200 years old, ranging in height from 60 to 140 feet. The three-seeded fruits are 23% to 32% oil. Sometimes this is expressed and used as a drying oil. The resultant residue, an oil cake rich in protein, is fed to livestock.

The tapping of rubber from wild trees in the Amazon is an inefficient system. First, trees must be located; then the bark is roughly cut, a procedure not conducive to continuing production. Also, there are no nearby processing centers, which are necessary for efficient mass handling of rubber. Most tappers in the Amazon smoke the latex over open fires on broad paddles—molding it into large balls, which they then carry long distances to market. The system cannot compete with that used in the Far East, where trained tappers extract latex from concentrated tree plantings next to processing locations.

A variety of means has been employed in the extraction of rubber. Trees may be tapped by slitting the bark, which cuts the lactiferous ducts; sometimes the entire tree is severed, and the latex obtained by girdling. This is the means of extraction used most often for wild species. Until the Malaysian plantations became a major rubber source, wild trees were being cut down and exploited by untrained tappers.

The *jebong* system of tapping originated on the Asian plantations. A spiral cut is made in the outer bark, about 6 feet from the ground. The cut is extended toward the ground as tapping proceeds. In this manner, trees are tapped every other day for nine months, with a three-month respite. Less vigorous trees are tapped every third or fourth day. A skillful tapper will do no lasting injury to the plant.

Rubber trees are vegetatively propagated, and frequently consist of three grafted sections: a desirable root system, a trunk with many ducts, a leafy crown that insures maximal rubber production. In this system a high-yielding clone is budded onto a native root when a plant is large enough; then it is grown in a nursery until the tree is six to ten feet tall. At that time a bushy top from a disease-resistant strain is grafted, and the tree is transplanted. Most cultivated rubber trees are productive for 12 to 25 years, and are replaced when the yield lessens.

Other Sources of Rubber   Other sources of rubber include Castilla, or Panama rubber, obtained primarily from *Castilla elastica*, a tree native to Mexico and Central America. Assam rubber comes from *Ficus elastica*, a tall, buttressed tree native to India and Malaysia. The Lagos silk rubber tree (*Funtumia elastica*), native to West Africa, was so wastefully exploited that it was nearly exterminated; but efforts are being made to reestablish it through cultivation. Landolphia rubber comes from various species of *Landolphia*, huge African vines growing to 6 inches in diameter, that yield a latex. Guayule rubber is obtained from a low, shrub-like plant (*Parthenium argentatum*) native to dry areas of Mexico and the southern U. S. There is no latex in the guayule; but there are small, scattered granules of *caoutchouc*. The rubber, if resinous materials are removed with solvents, is as good as *Hevea* rubber, and is especially useful for mixing with synthetic rubbers. Latex can also be extracted from tubes in the long tap roots of Russian dandelions (*Taraxacum kok-saghyz*). Cassava (*Manihot*) yields a good rubber, and is cultivated in Brazil for that purpose.

Chicle, the basis for chewing gum, is obtained from the coagulated latex, *balata*, from the sapodilla tree, *Achras zapota*. Balata from *Achras* is extracted by gashing the standing tree from top to base. The latex is often collected by small groups of individuals who scour the jungle—collecting as much as 60 pounds from a single tree. Another natural balata is *Palaquium*, which yields *gutta-percha*, a nonelastic rubber of Malaysian origin. It is a popular insulating material. *Chilta*, from a species of *Anidoscolus* which grows in dry regions of Latin America, provides a nonelastic rubber used as an extender for more useful elastomers, such as gutta-percha. The genus *Couma*, from which gutta gum is obtained, grows throughout northern South America. High in resins, gutta gum is used in the U. S. as a component of chewing gums. Locally, it serves as a caulking compound, and when fresh is sometimes used in place of cream in coffee.

Suggested Readings

Aykroyd, W. R. *The Story of Sugar.* Chicago: Quadrangle Books, 1967.

Dijkman, M. J. *Hevea—Thirty Years of Research in the Far East.* Florida: University of Miami Press, 1951.

Great Western Sugar Company. *How to Grow Sugar Beets.* Denver, Colorado: Through the Leaves Press, 1951.

Haves, F. N. *Vegetable Gums and Resins.* Massachusetts: Chronica Botanica, 1949.

Hill, Albert F. *Economic Botany.* New York: McGraw-Hill, 1952.

Jenkins, G. H. *Introduction to Sugar Cane Technology.* New York: Elsevier Publishing Company, 1966.

Paturau, J. Maurice. *By-Products of the Sugar Cane Industry.* New York: Elsevier Publishing Company, 1969.

Polhamus, L. G. *Rubber: Botany, Production, and Utilization.* New York: Interscience (Wiley), 1962.

Stubbs, William C. *Sugar Cane.* Louisiana State Bureau of Agriculture and Immigration, 1904. (This book provides a historical perspective for those interested in early sugar cane cultivation in the United States).

United States Sugar Beet Association. *The Sugar Beet Story.* Washington, D.C.: The United States Sugar Beet Association, 1959.

Weyl, Nathaniel (1970). "Some Genetic Aspects of Plantation Slavery." In *Genetics and Society,* edited by J. Bresler. Reading, Mass.: Addison-Wesley, 1973.

CHAPTER
11
THE LONG ROMANCE OF
SPICES AND PERFUMES

Essential oils are highly aromatic substances which impart flavor to spices and other natural flavorings, give fragrance to perfumes, and add the "clean" odor to antiseptics and medicinals. One distinguishing characteristic is that they do not leave a "greasy" spot on paper. These essential oils are mainly benzene and terpene derivatives. A typical molecule is seldom more than 20 carbon atoms long. They turn readily into resins and gums when allowed to oxidize in air or light. Typically, the liquids evaporate easily, a quality known as volatilization.

Historical Significance

Although basically irrelevant to human welfare, essential oils have featured prominently in the history of Western civilization. Expeditions of European explorers were launched as often in search of spices and perfumes as in the pursuit of gold and other valuables. The news that the Far East was abundant in these materials served as the impetus for voyages to colonize spice-rich lands. Europeans were not the only cultivators and users of spices. Perfumes and spices have been important to many societies since the beginnings of recorded history. In ancient China, India, Babylon, Egypt, Greece, and Rome spices were held in esteem. Spices were among the first items to inspire trade between the East and the West. Arabs were the first traders, bringing spices from India and the Spice Islands to Arabia by caravan, and from there to Europe. In Europe Venice then became a spice center, followed by Portugal, which retained dominance for 200 years. Next the Dutch gained control of this lucrative market, a control later shared with Britain. Today, spices are so effectively distributed throughout the world that the significance of the Far East as a spice center has considerably diminished.

The Role of Spices Today

In affluent countries we take spices for granted, so it is difficult to comprehend the seemingly exaggerated importance of spices in earlier times. In the past the human diet was far more monotonous than it is today in developed nations. Human beings ate a few staple items. Spices provided the means for varying the somewhat drab and often insipid fare, and also acted as preservatives, a significant feature when there was no refrigeration besides the outdoor temperature.

Perfumes, although still the basis of a multimillion dollar business, were perhaps of more direct importance when people were less conscious of cleanliness. During cold European winters, baths were an infrequent luxury. The odors of humanity must have been pleasantly camouflaged by the aromas of various perfumes.

Although natural spices are currently not sought as in past centuries, millions of dollars are spent in the United States for spices on the wholesale market. Most companies import crude spices; then prepare the spice for commercial use in the United States. This prevents possible adulteration. Spice quality is strictly regulated by the Pure Food and Drug Administration.

Spices cannot really be considered food. The nutritional value is nil, but the role of spices in enhancing the flavor of foods should not be minimized. Many processed foods are made palatable by either natural or synthetic flavorings. Most prepared foods contain some kind of flavor additive. During the Middle Ages, spices were considered very valuable as medicines. In modern cultures, spices are still used to relieve colic, as antiseptics, and to mask the disagreeable tastes of some drugs.

Essential Oils and Spices of Major Importance Most essential oils come from tropical plants, and most of the commercial yield still comes from Asia, with Africa and tropical America supplying the balance. The classification of spices is arbitrary, as spices are derived from a variety of plants. "Spices" generally include all plant products used to flavor food and beverages. Condiments, spices having a sharp taste, are usually added to food in small amounts after cooking. Savory seeds are sometimes small dried fruits or seeds, such as sesame, that are sprinkled on food without being ground. Sweet and savory herb leaves are used fresh or dried. Essences are aqueous or alcoholic extractions of essential oils. There are many hundreds of substances used in spices and perfumes, and to discuss them all would require a booklength text. Here we are going to review only a select few of major importance.

Black and White Peppers. These come from East Indian pepper plants (*Piper nigrum*). They are perennial climbing vines; the hard berries (peppercorns) are dried or naturally fermented. The white pepper is produced by immersing the berries in water for several weeks. After soaking, the berries are allowed to ferment naturally, and then the outer hull is removed. Black pepper is obtained by quickly drying the entire fruit. The plants are usually propagated through cuttings. Annual world production is about 100,000 tons.

Red cayenne pepper (Capsicum spp.), a hot and pungent powder, is made by drying and grinding an American chili. Other chilies, paprikas, and sweet bell peppers are also members of this genus, and represent the United States' chief spice contribution.

Cinnamon *(Cinnamomum zeylanicum).* A plant native to Ceylon and India, cinnamon was one of the first spices used by humans. It has been known for 5000 years. The spice is bark cut from sucker shoots that develop on the roots. The trees are propagated by cuttings. Cinnamon is a popular spice used today in candies, incenses, gum, dentifrices, and perfumes.

Cloves. The unopened flower buds of *Eugenia caryophyllus,* this is one of the more widely used of all spices. The plant is native to a single island of the Moluccas, yet by 2300 BC it was already widespread as a spice and as a breath sweetener in China. Today, cloves are grown principally in Madagascar and Indonesia, and to a lesser extent in other tropical countries of the New and Old World.

Vanilla. This flavoring is extracted from the ripened seed pod of a climbing orchid (*Vanilla planifolia*) indigenous to moist forests of Central America. The pods are cured by exposure to the sun during the morning, then protected by blankets or other cover in the afternoon, and placed in airtight boxes at night. During this process, enzyme action transforms a glucoside into a crystalline substance *vanillin,* which has the characteristic odor and flavor we know. Now the fragrance and flavor of natural vanilla can be easily synthesized, the artificial product being similar to the natural one. The Maya and the Aztecs used vanilla as a flavoring in chocolate; we still do this today. The horticulture of vanilla is expensive because the plants must be pollinated by hand.

Mints. These plants (*Mentha spp.*) have been used culturally for thousands of years. Peppermint and spearmint were known to the Egyptian pharaohs. The former is one of the most important of aromatic herbs, having a persistent "cooling" taste and a refreshing odor. The commercial production of mint oils in the United States is more than 100 tons per year. Mints are planted in rows as cuttings and permitted to grow for the first year. The plants are then plowed under during the first fall—to revive the following year as meadows of mint, at which time the plants are ready for harvesting. Mint gives flavor to peppermints, mint jellies, and other foods. The oil, used in cosmetics, food, and medicines, is expressed by steam distillation. Waste material from the process is often returned to the soil as mulch.

Oregano. Originally a spice from a savory herb (*Oreganum vulgare*), today the spice is sometimes commercially derived from the dried leaves of other plants as well. Native to the Mediterranean, where it is a much-esteemed flavoring, oregano was brought to the United States by immigrants. *O. vulgare* and related herbs were common in pre-Renaissance European cookery, before other spices became reasonably abundant.

Spices from Roots. Spices obtained from roots or rhizomes include ginger (*Zingiber officinale*), horseradish (*Rorippa Armoracia*), sarsaparilla (*Smilax spp.*) and turmeric (*Curcuma longa*). Of these, ginger is the most important. Originating in southeast Asia, it had been carried by caravan to Asia Minor before the existence of Rome. During the reign of Henry VIII, it served as the basic ingredient in a plague remedy. *Z. officinale* is currently cultivated in both the New and Old World tropics, with India and Taiwan supplying three-fourths of the commercial yield. The ginger spice comes from the ground rhizomes. It is a widely used culinary ingredient in the tropics, in part because it causes blood vessels in the skin to dilate, increasing perspiration and an accompanying drop in body temperature. Horseradish roots are scraped or grated, and are used chiefly as a condiment. The strong, pungent taste is due to the presence of a glucoside, *sinigrin*. Sarsaparilla, the dried root of several tropical American trailing vines, is a flavoring usually found in combination with wintergreen and other aromatics. Turmeric, related to ginger, yields both a dye and a spice. It is a principal ingredient of curry powder, and is cultivated extensively in India.

Mustard. A spice obtained from the tiny seeds of two species of *Brassica* native to Eurasia, mustard has been used commonly since Biblical times. Although the seeds are by far the most important part of the plant, the green tops are used in salads and other dishes. The plants cultivated for seed are white mustards (*Brassica alba*) and black mustards (*Brassica nigra*). The small, round yellow seeds of the former grow on a branching annual that may be from 2 to 6 feet tall. The seeds contain mucilage, proteins, oil, and *sinalbin*, a glucoside. When the seeds are ground and mixed with water, sinalbin is broken down by enzymes to yield a sulfur compound responsible for the sharp, pungent flavor. Black mustard is more abundant, and has smaller seeds than the white mustard. The latter has smooth pods containing dark brown seeds which are yellow inside. Both black and white mustard seeds have approximately the same components. The sinalbin, if very diluted, can be used as a condiment or as a counterirritant.

Above *Cinchona bark
is removed from a
seven year old tree cut
from a Guatemalan
plantation. (USDA
photo)*

Left *Pine trees in
Honduras slashed for
resin collection. (FAO
photo)*

Perfumes, Cosmetics, and Soaps The essential oils basic to the manufacture of perfumes are usually obtained from flowers. Those used to give agreeable scents to soaps and cosmetics are usually extracted from leaves. Flowers have nectar glands which secrete essences to lure insects for pollination (although a flower's fragrance may not be due to nectar). The leaves of the same plants will often have epidermal glands or hairs that produce essential oils.

Historical Significance   Perfumes and related substances have been of importance to human beings for millennia. In ancient civilizations, perfumes were an expression of animistic and cosmic conceptions, supposedly a means for eliciting divine responses to a society's problems. Even today, scents, primarily in the form of incense, are an integral part of many religious rituals. In China, perfumes have been used for thousands of years, and in India there were ancient rituals involving perfume. Buddhist liturgy decreed that statues of gods must be washed in perfumes, and similar customs were practiced in ancient Egypt. In the five books upon which the history of Israel is founded, there are numerous references to perfumes. During the height of Grecian civilization, perfumes were used in many ways. Hippocrates described some of these, including *Iatralypte,* a term derived from the physician who cured by using perfumed unguents. Pliny the Elder wrote about a process for obtaining perfumes from fats, and observed that the Romans deemed the use of perfume to be one of the most "honest" of human pleasures. The Arabs were also greatly interested in perfumes; the *Antidotaire* of Mesuë, a twelve-book pharmacological treatise, included material on perfumes. In Europe, perfumes as well as spices instigated the exploration and colonization of parts of the New World and the Far East. During the 18th century, the people of Provence in France produced most of the perfumes of consequence, except for a few from Italy.

Extraction of Essential Oils   Essential oils are absorbed readily into cold fat of any kind—ordinary lard works best. Glass sheets are coated with the fat, and petals or leaves are laid upon the coating. Generally, the petals are changed from two to ten times, until the fat is sufficiently aromatic. This practise is known as *enfleurage,* a method employed widely in the Provence perfume center of France. Afterwards, the essential oil can be separated from the lard medium by alcohol extraction. Stems and leaves are often boiled, a procedure that releases the oils, which then float to the surface. Also, some flowers may be "digested" at temperatures ranging from 40° to 70°C in a mixture of fats or in oils. *Flower pomades* are the result of this procedure. Perfumes may be compounded from as many as 20 of these oils, and are also made from animal oils, such as various musks and ambergris, a grey-white, waxy material produced by sperm whales. The latter was especially prized in former times, and is still employed in costly perfumes. Now there are also many synthetics available which are used in the creation of a variety of inexpensive perfumes.

The alcoholic extraction of essential oils from fats results in *tinctures* if heat is used during the process. *Infusions* are produced if heat is not used during the process. Alcoholic extraction is not effective for obtaining oils from fresh flowers or leaves. The water in the plant dilutes the alcohol, so it loses its solvent capacity. When the isolation of fragrances from fresh material is necessary, ether or other volatile solvents, such as benzene, are used. Extracts obtained from dry plants are rich in resins; those from fresh plant parts are high in waxes and are often solid, or nearly so.

Rose Oil. This is a popular essence produced predominantly in southern France and Bulgaria. Less than a half-gram of oil is obtained from 1000 grams of flowers. Rose water, a frequent ingredient of Greek and Middle Eastern pastries, is a by-product of this industry. The oil, sometimes called *attar* or *otto* of roses, is often adulterated with geranium or palmarosa oil. Both are less expensive oils with a roselike odor that blends well with the real substance.

Patchouli. This essence is extracted from the fleshy leaves and young buds of *Pogostemon cablin*, a mint native to the Philippines. It is unknown how long the plant has been grown in the Orient. Currently, *P. cablin* is raised commercially in Indonesia, the Seychelles Islands, China, and southeast Asia. It is propagated by cuttings. The leaves are cut and then left to ferment naturally in heaps. After fermentation, the leaves are distilled to yield a dark brown oil with a strong odor reminiscent of sandalwood. The oil constitutes 3% to 5% of the dry weight of the leaf. Originally, patchouli was used as an insecticide. Today, it is an ingredient of soaps, hair tonics, and tobacco, and is one of the better fixatives for heavy perfumes. Patchouli also produces the characteristic odor of cashmere shawls, which are often shipped in containers scented with the fragrance.

Camphor Oil. This is perhaps the most important of essential oils for industrial purposes. It is extracted from the camphor tree (*Cinnamomum camphora*) of Japan and China.

The wood is used for chests because the oil will repel insects. Camphor trees were threatened with extinction because so many were cut for the oil. Now only trees 50 years old, or more, are used. To extract the oil, the wood and leaves are ground or chipped; then distilled with steam for about three hours. During that time, crude camphor crystallizes on the wall of the still. Camphor is used in perfumes, medicines, and in the manufacture of celluloid and nitrocellulose compounds.

Other Important Oils. Other oils are obtained from *Citronella* grass (*Cymbopogon nardus*), cultivated in Java and Ceylon. Relatively inexpensive, it is found in many soaps and perfumes and is an effective insect repellent. It is also used in the synthesis of vitamin A. *Sandalwood* oil, from the wood of *Santalum album* and some related species, is popular in the Far East as a perfume and an ingredient of medicines. The trees grow wild in India and other parts of southeast Asia, and are cultivated in many countries. In some areas where sandalwood is indigenous, the trees have nearly been eliminated by exploitive lumbering. Sandalwood chips are sometimes placed in containers for their fragrance; the wood is often used in making cabinets and chests. In India, it is placed on the funeral pyres of the wealthy. *Carnations* (*Dianthus caryophyllus*), of which there are more than 2000 varieties, supply an oil having a rich, sweet odor. Lavender, from the south European shrub *Lavandula officinalis*, is an important scent in high-quality perfumes, soaps, and cosmetics. It is also a mild stimulant used in medicines. *Ylang-ylang*, meaning "flower of flowers," is an Asiatic tree (*Cananga odorata*) which yields one of the most valuable of perfume oils, derived from its opened flowers. Although expensive, it is still in demand. *Jasmine*, from *Jasminum officinarum*, var. *grandiflorum*, is a highly regarded source of perfumes. The oil is isolated by solvent extraction as soon as possible after the flowers have been picked. *Orange blossom* oil, from the flowers of *Citrus aurantium*, is significant primarily as an extender of other, more valuable oils, and as the basis for cheap perfumes. *Geranium* oil, from the foliage of several species of cultivated geraniums (*Pelargonium spp.*) is commonly used as an extender or as a blend with rose oil.

Suggested Readings

Arctander, S. *Perfumes and Flavor Materials of Natural Origin* (2 parts). Westport, Connecticut: Avi Publishing Company, 1960.

Guenther, E. *The Essential Oils*. Princeton, New Jersey: Van Nostrand, 1948.

Merory, Joseph. *Food Flavorings*. New York: Chemical Publishing Company, 1968.

Naves, Y. R., and Mazuyer, G. Translated by Edward Sagarin. *Natural Perfume Materials*. New York: Reinhold Publishing Company, 1947.

Parry, J. W. *Spices*. New York: Chemical Publishing Company, 1962.

Rosengarten, Frederic. *The Book of Spices*. Wynnewood, Pa.: Livingston Publishing Company, 1969.

CHAPTER

12

BEVERAGES

Above *Coffee beans are still picked by hand. The berries grow in clusters along the stem and mature at different times. (FAO photo)*

Left *Workers sorting coffee beans at a cleaning plant in Addis Ababa, Ethiopia. Ninety percent of the country's population of approximately 24 million are employed in agriculture or agriculture-related jobs. (FAO photo)*

Liquids are essential to human life. For as long as people have been searching for ways to diversify their existence, they have surely been experimenting with variety in beverages. Most of the culturally important drinks, with the exception of milk, are derived from plants. Thousands of species have been used to make beverages, but relatively few have economic significance.

Generally, beverages may be divided into those that contain alcohol and those that do not. Of the former, beer, wine, and distilled spirits, such as whiskey and rum, predominate.

Nonalcoholic Beverages Of the nonalcoholic beverages, almost all have a stimulating effect, due to the caffeine or other alkaloid content. Many artificial fruit drinks and colas are infused with stimulants, which increases the drink's popularity. Although caffeine has been associated with coronary attacks, the evidence is tenuous. It is probable that a healthy adult can drink coffee, tea, and similar beverages in moderation with no ill effects.

Coffees (*Coffea spp.*) Domesticates of relatively recent origin, coffees are today the most important beverage plant economically—although more tea is probably consumed. Coffee is drunk regularly in a third of the world, with the United States importing the greatest quantity. Native to Ethiopia, coffee was cultivated in Yemen by 1500 AD, and Europeans were familiar with the beverage by then.

It was discovered that coffee plants grew well in the American tropics, from elevations near sea level to more than 6000 feet. Brazil is currently the foremost producer, followed by Colombia, parts of Africa, and Central America. World production is more than four million tons per year. The principal difficulty in cultivating coffee arises from a leaf rust that decimates crops every few years, causing severe cuts in production. The plants grow well in shaded forest. On commercial plantations, coffee is planted as an understory crop.

The Blue Mountain coffees of Jamaica are one of the more flavorful varieties, a quality apparently due to the environment rather than to the plants. Plants transported to Africa from Jamaica yielded no more flavorful coffee than varieties already under cultivation in Africa. Congo, or Robusta coffee, is a large, vigorous plant with a high yield and a strong resistance to leaf rust. Colombian and Central American coffees which are *c. arabica* seem to have a better flavor than do African types.

On maturity, coffee berries are picked by hand. Each fruit contains two seeds. These are the coffee "beans," from which

Above *Cleaning dried tea leaves in a factory in Kandy, Sri Lanka. (FAO photo)*

Right *A sprig from the tea plant. (FAO photo)*

the pulpy pericarp is removed through natural fermentation. The beans are usually dried by spreading them out under the sun. Roasting, the process that gives coffee its characteristic taste, is done just before packing, as the flavor deteriorates after roasting and grinding.

Tea

An important drink for perhaps half of the world's population, tea comes from a broad-leaved evergreen, *Camellia* (or *Thea*) *sinensis*. The evergreen leaves provide the beverage. The plant is native to India and China; tea has long been used in the Orient as a medicinal and a beverage. It was first imported into Europe by the Dutch in 1610; and was taken to London in 1664; then Boston in 1714. In China and Mongolia, tea leaves are extracted from the drink, mixed with butter and salt, and eaten. This same custom was also practised in Salem, Massachusetts. Tea was introduced into Japan about 1000 years ago, and today is a drink of major importance there. In Britain, its popularity superseded that of coffee. Now, tea is as socially important in England as anywhere in the world.

In nature, the plants may reach a height of 30 feet. Under cultivation, they are kept to the size of a shrub—usually 3 or 4 feet tall. The nature of the plants and the means of cultivation vary from one locale to another, as there are more than 1000 varieties. Yields range from 200 to 1000 pounds per acre. Plants are long-lived and a single plant may be productive for 40 or 50 years. Some plants in Japan have lived for more than two centuries. Tea plants will grow from sea level to 5000 feet. They will thrive in relatively poor soils, but prefer a moist environment with a rainfall of no less than 60 inches and a temperature range of 70° to 90°F. Harvesting is by hand, which accounts for tea being grown predominantly where manual labor is not expensive.

After the leaves are picked, they are heated by the sun or over fires in small trays until pliable. At this point, the leaves are rolled, by hand or machine, to curl them and remove some of the sap. Green teas are made by rapidly drying the leaves. Black teas, the kind most popular in Europe and America, are made from the same plant species by fermenting the leaves, then rolling or trampling them to break the cells and allow further fermentation. During this process, the black coloration and characteristic flavor is developed. The final product provides a stimulating drink, for tea contains more caffeine than coffee. It also has several parts per million of fluoride, and tannin, which are both agents in retarding tooth decay. Tea drinks may be brewed from many herbs lacking caffeine. Their use has become increasingly popular.

Theobroma cacao tree with mature cauline fruit. *(USDA photo)*

Cocoa

When the Spaniards conquered Mexico in 1519, the reigning Aztec emperor was receiving an annual tribute of 3000 tons of *cacao* beans, the seeds from the tropical American tree *Theobroma cacao*. The term "chocolate" is derived from the Aztec term *Xocoatl*, or *chocoatl*. Although cacao trees were grown in Central America for centuries before the European invasion, their origin has been traced to the Amazon Basin. First domesticated by the Mayas, the trees were eventually transported to West Africa by the Portuguese. It is here that most cacao is now grown. Ghana and Nigeria cultivate about a third of the entire commercial crop; other parts of Africa and Central America supply the rest. Due to widespread cultivation, the trees now grow wild in many tropical forests.

In nature the cacao tree is small and thrives in the understories of tropical forests. Commercially grown trees are cultivated in the shade of other trees, because the young cacao plants burn in the tropic sun.

The flowers and fruits are *cauline*, growing on the lower stems of the plant. The football-shaped fruit pod ranges in color from red to yellow on maturity, depending on the variety. The cacao beans number 20 to 40 per pod. To harvest, pods are hand collected, split open with machetes, and the beans scooped from the pulpy center and placed in boxes or baskets to ferment. Fermentation changes the color of the beans from ivory to purple, and brings out the characteristic chocolate flavor and aroma in the seeds.

When dried, the seeds are shipped to processing factories where the procedure for chocolate manufacture begins. Beans are graded and blended, in much the same manner as coffee, to produce the desired color and flavor. Following grading, they are roasted, and the meat of the seeds is extracted mechanically. The nut meat, which contains 53% cocoa butter and forms the basis for all chocolate products, is ground and pureed in its own oil to form a liquid. At this stage, the manufacturer makes commercial cocoa by removing most of the cocoa butter, or makes chocolate by adding more cocoa butter. Unsweetened cooking chocolate is simply liquid cocoa butter cooled and molded. The natural product is bitter, so oil and sugar are added to the chocolate used for confections.

Chocolate is rich in nutrients—30% to 50% oil, 15% carbohydrate, and 15% protein. The plant's name, *Theobroma*, means "food of the Gods" in Greek. Cocoa, a drink made by mixing powdered chocolate and sugar in boiling water or milk, is especially popular as a breakfast drink.

In Mexico and Central America, one may purchase a variety

of instant cocoa beverages which are simply crude cocoa with sugar added. Coarse and grainy, the mixture is sometimes eaten as candy, but is better as the ingredient for hot chocolate. The beverage has probably been consumed for hundreds of years. The Spanish reportedly kept the technique for chocolate production a secret for as long as they could. Finally it became known, and the great chocolate houses of Cadbury in England and Van Houten in Holland were established; they still exist today.

Maté   This is a South American drink derived from the leaves of a holly *Ilex paraguariensis*. It has little commercial significance, but is consumed by more than 20 million people. The plants grow wild in Paraguay, Argentina, and the mountains of southern Brazil, and are also, to some extent, cultivated. A small quantity of maté is imported into Europe and North America. The beverage originated with the Guarani Indians in South America. It was adopted by European settlers, and Jesuit missionaries encouraged the planting of the trees. The leaves are dried, crushed, and brewed in the same manner as tea leaves. Traditionally, maté is drunk from a gourd through a metal straw.

Other Beverage Plants   These include *guarana*, made from the seeds of a large, woody climber (*Paullinia cupana*) of the Amazon valley. The caffeine content of its seeds is three times that of coffee. *Khat*, obtained from the leaves of a shrub, *Catha edulis*, is a native drink in Arab countries. *Cola*, a carbonated beverage additive from the seeds of *Cola nitida*, is used in beverages in Africa, where the plant grows, and in beverages manufactured in many other parts of the world. The seeds are also chewed, as a stimulant and mild pain-killer. *Yoco* plants (*Paullinia yoco*) yield a beverage from the bark. Extracted in cold water, the drink contains from 3 to 6 percent caffeine. Some groups of Indians in the area of southern Colombia, Peru, and Ecuador will abandon otherwise amenable village sites if nearby supplies of the plant become scarce.

Alcoholic Beverages   Undistilled kinds of alcoholic beverages have been brewed for many thousands of years (Rose, 1959). There are depictions of beer brewing on Sumerian and Assyrian clay tablets more than 6000 years old. And there is evidence that prehistoric humans were wine makers.

Beer   When Columbus arrived in the New World, he was introduced to an Indian beer, *chicha*, made from maize. During the Middle Ages, beer was drunk throughout Europe, and the brewing of beer was a routine household responsibility. There was no refrigeration, so it was sometimes lagered in the coolness of caves. Monasteries brewed great quantities, and their influence is still seen in the symbols of quality XX and XXX. Beer was also one of the first industrialized products. By the mid-1800's, several major brewing centers existed; Oxford, Burton-on-Trent, Munich, Milwaukee, and Pilsen. As a source of government revenue, a tax on alcohol is not new; Charles I of England created the first beer tax in 1643.

What we consider beer today is usually a brew of barley malt, made bitter by the use of hops. A key step in beer brewing is the input of yeast, which breaks down grain sugars into alcohol and carbon dioxide. This results in "green" beer. The nature of the grain used is another essential factor. Barley is perhaps the most significant grain . The word "beer" is derived from *baere,* the Saxon word for barley. The food reserves of a barley seed—especially the polysaccharides—are of major interest to a brewer. Conventional beer is made of malt, which is obtained by germinating the seeds of the barley (or other cereal). Germination converts the starch of the grain into sugar, which then can be worked upon by the yeast. Germination is stopped by drying the grain seeds in a kiln, at 180°F. Higher temperatures are used to caramelize the barley if dark beer is desired. The barley is then added to a starchy material. This mixture of barley and the adjunct is the *mash,* which is soaked in hot water to produce *wort.* The wort is then boiled with hops to achieve the characteristic semi-bitter taste.

Hops (*Humulus lupulus*) are added for flavor. In the Middle Ages, herbs were used to perform the same function. Brewers in the U. S. usually employ about a pound of hops per barrel. European brewers may use as much as four pounds per barrel (a barrel holds 31 gallons).

Hops are cone-shaped inflorescenses of female plants. The hops are picked when ripe, dried, and then shipped to breweries. Most hops in the United States are grown along the Pacific Coast and in Idaho. The hops are boiled to extract a variety of substances such as tannins, oils, and complex bittering materials. Antiseptics also are drawn out and aid in the prevention of microbial spoilage. The tannins are valuable in brewing—combining with protein in the wort to form an insoluble sludge which settles to the bottom of the brew kettle. Were the protein to remain in the wort, it would precipitate later as a haze, and detract from the beer's appearance.

When the wort is cooled, it is piped into huge tanks. Here the mixture is inoculated with a particular yeast (*Sac-*

charomyces cerevisiae), which ferments the wort—yielding alcohol, carbon dioxide, and some minor organic compounds which impart to the beer its customary taste. After only a day or two in the starting containers where the yeast is injected, the beer is transferred to lagering or aging vats. Ideally, it is kept for weeks or even months in these vats, becoming carbonated with the buildup of carbon dioxide. However, most commercial beer is artificially carbonated to save time. In either case, a small amount of green beer may be introduced directly before the beverage is bottled or kegged. This gives a last-minute boost to fermentation. The finished beer is about 90% water, 5% ethyl alcohol, and small amounts of maltose, gums, dextrins, and nitrogenous materials.

Lager beer refers to beer that has been aged for a longer time. Bock beer is a very dark, strong variety made from the first of the new malt and hops. Ale, fermented by floating colonies of yeast, usually has a higher hop and alcohol content than beer. It is also brewed at a higher temperature. Porter is a dark-brown, sweetish beer made from caramelized malt. Stout is similar to porter, but heavier, with a higher percentage of alcohol. Weiss, made from wheat, is a light and malty ale.

Wine

Wine is a mildly alcoholic beverage consisting of ethyl alcohol, several other alcohols, sugars and other carbohydrates, polyphenols, aldehydes, ketones, enzymes, pigments, six or more vitamins, 15 to 20 minerals, and more than 22 organic acids. There are many possible variations in the formula, and there are a number of varieties and qualities of wines. The processes for making wines are now well known, so that the production of a good wine can be fairly certain. The qualities characterizing a "great" wine, however, are elusive (Amerine, 1964).

People have enjoyed wines for a very long time, just how long is not certain. We know that wine has been of importance to various cultures for thousands of years. Wine's "spiritual" origin may be Europe, or perhaps the Near East. Today, it is produced in North Africa, South Africa, Chile, Argentina, Australia, and in the United States, particularly in California and New York. The wine-making process is relatively simple, so it was probably developed independently by different peoples at more than one time.

As with beer, wine is fermented by the action of a yeast, Saccharomyces ellipsoideus, the so-called wine-yeast. On the skin of any given grape, there may be as many as 10 million yeast cells; of these, perhaps 100,000 will be wine yeast. It is these cells which will cause fermentation if the grapes are converted into wine. The quality of a wine depends on the nature of the actions of the yeast, a factor carefully controlled in wineries. Most modern wine producers add pure cultures of desirable yeasts and certain chemicals, such as sulphur dioxide, to inhibit the proliferation of undesirable kinds of yeast. The basic principle of wine-making is uncomplicated. The grapes are crushed into a *must*, which is allowed to ferment naturally. When the alcohol content is from 10% to 14%, the yeast is killed by the alcohol. "Fortified" wines are those to which alcohol has been added. An example is sherry, which has an alcohol content of 20%.

"Wine" usually refers to the fermented product of a particular grape, *Vitis vinifera*, (the European grape), but wines can be made from almost any kind of fruit, simply by allowing the fruit to ferment. The nature of a wine is directly related to the conditions under which the grapes are grown. Generally, an "ideal" environment for wine grapes is one of moderate coolness and moderate warmth, thus the success of vineyards in California. A long growing season is desirable so the grapes can build up stores of sugar. Coolness is favorable, as it results in grapes having a higher acidity. This is an important characteristic in producing drier table wines.

Wines can be separated into "beverage wines," a category including about 95% of the wine produced, and "fine wines." The beverage kinds, also known as *vins ordinaires*, are not aged to any degree. The fine wines are more expensive, and frequently have been aged for at least several years before being marketed. *Red wines*, which tend to be heartier, are made from grapes with skins that contribute to the natural coloration. *White wines* are made from white grapes or from expressed juice. "Sparkling" wines are bottled before fermentation is complete so that some of the carbon dioxide is captured in the bottle. Wines are about 70% water, and contain from 10% to 14% alcohol. The acidity ranges from essentially zero to 0.65%. Some organic compounds are present, as are various trace minerals which may be beneficial to human health. *Vermouths* are wines to which bitters (in particular the leaves or shoots of *Artemisia spp.*) have been added.

Distilled Spirits Natural fermentation can yield beverages having an alcoholic content of no more than 10% to 14%, because alcohol in higher concentrations kills the yeast. Wines or beers are *distilled* to yield spirits with a much higher content of alcohol—usually about 50%. Alcohol has a much lower boiling point than water; so, during distillation the alcohol is vaporized and collected in condensers. The basic process is simple, although to produce distilled spirits of quality one must be

The Maguey (agave) plant with center removed for the extraction of juice. The juice is fermented to make "pulque," a popular drink in rural Mexico. (FAO photo)

skilled, because very exact blends of ingredients are required to achieve characteristic flavors.

A mixture of highly distilled beer and water yields *whiskey*. The more expensive kinds are aged in charred-oak containers for at least four years, and often longer. Whiskeys are made either from corn, barley, or rye. Bourbon whiskey, originally produced in Bourbon County, Kentucky, is made predominantly from corn, the rest from rye. *Brandies* are spirits made from distilled fruit wines; they are high in alcohol content, usually 65% to 70%. *Rum* was originated in tropical countries and is a distillation of the juice, molasses, or other unrefined products of sugar cane. *Gin* is a distillation of fermented mash, the more expensive are made from barley mash and rye. The characteristic flavor of gin comes from Juniper "berries" used in the distillation process. *Tequila* is a distillation of fermented juice from leaf bases of *maguey* plants.

Suggested Readings

Amerine, Maynard A. "Wine." *Scientific American* 211 (1964): 46–56.

Amerine, Maynard A., and Singleton, V. L. *Wine.* Berkeley: University of California Press, 1968.

Haarer, A. E. *Modern Coffee Production.* London: Leonard Hill, Ltd., 1962.

Harler, C. R. *Tea Growing.* London: Oxford University Press, 1966.

Hill, Albert F. *Economic Botany.* New York: McGraw-Hill Book Company, Inc., 1952.

Rose, Anthony H. "Beer." *Scientific American* 211 (1959): 46–56.

Schery, Robert W. *Plants for Man.* 2nd ed. Englewood Cliffs, N.J.: Prentice Hall, 1972.

Stanier, Roger Y.; Doudoroff, M.; and Adelberg, E. A. *The Microbial World.* Englewood Cliffs, N.J.: Prentice Hall, 1970.

Wagner, Philip. "Wines, Grape Vines, and Climate." *Scientific American* 230 (1974): 106–114.

In Unit Four, we consider *ethnobotany* to be the interactions between human beings and plants as discrete cultural entities. We examine the role of plants in human cultures from a somewhat broader perspective than that of most "economic botany" texts. The information contained in this unit will give you a more comprehensive appreciation of the various uses of plants as food, as medicinals, and as religious vehicles.

UNIT 4

CHAPTER

13

METHODS IN ETHNOBOTANICAL INVESTIGATION

In ethnobotanical studies, the fields of botany and anthropology merge. The ethnobotanist attempts to accomplish two things: to discover which plants are commonly used by a people, and to determine for what purposes the plants are used. Local identification and scientific identification of plant materials is necessary. The ethnobotanist makes collections of plants and local plant products as scientific specimens for later identification. The traditional local uses of plants must be verified. In these areas, a general knowledge of field anthropology is necessary, as well as a background in botany.

Consider some of the problems in logistics and interactions that a field worker investigating the ethnobotany of an area might encounter:

Selection of a Site Most ethnobotanical studies have been conducted in one or a few relatively small communities of a wider culture. There are well-founded, practical reasons for this, as will be discussed. Even if a broader base of operations is needed for the collection of materials from differing habitats, the ethnobotanist will often select one local community as a base for operations. This choice clearly defines the research area. The problem of transportation is solved if studies are restricted to smaller areas. Also, establishing the investigator as a resident in a particular community is important—not only for purposes of securing information critical to the study, but to give him or her a living base during the on-site period.

Ideally, living headquarters should typify local customs and other aspects of the cultural area. The site should be remote from the influences of larger polycultural groups. Transcultural influences diminish the importance of local customs and beliefs which reinforce the social continuity of folklore. With modern communication and transportation, such areas of the earth are becoming scarcer. It is to the advantage of the researcher and the results of the study if the community is at least reasonably cooperative and accepting of the stranger's presence. Language should not be a barrier to communication, community interaction, information collection, and the understanding of innuendos. A researcher's mastery of dialects may be particularly important, as many persons involved in the collection and use of plants as medicinals or magicals may be on the fringes of the established culture.

Impression Management A worker introduces himself or herself to the society of the study area as an outsider in almost all respects. If the fullest information is to be gained, it is important to avoid alienation of any significant sources within the community. As a primary consideration, the researcher's housing should be in a socially neutral location. This location may not be easy to decide upon immediately, and may be clear only in hindsight. For example, to be housed in government quarters, in a church rectory, or in housing provided by local political or religious leaders, would probably hinder communication with many of the people who could offer the best data. Folk medicine and shamanistic practises are frequently condemned by more socially and politically powerful members of a community.

The social role of outsider can work to the researcher's advantage. Since he or she will be known as a transient in the community, if an objective neutrality is maintained, the investigator may be the recipient of information which would otherwise be unavailable even to prestigious community members. The social role of interrogator is special. If a worker is careful to remain aloof to community frictions, there should be little difficulty in remaining neutral, and in a position to collect much significant data.

An ethnobotanist, especially if he or she is engaged in field work in a semi-urbanized area, must concentrate on acquiring a reputation as a person who does not betray confidences. There may be social stigmas associated with shamanism and folk medicine. Also, there may be religious or legal sanctions against it, or both. In situations where this is the case, it may prove nearly impossible to secure information of value. Nevertheless, if trust and objectivity can be maintained, an ethnobotanist will often gather information that could never be acquired were he or she not alien to the culture. However, if a researcher gives the persons who are information sources reason to believe that he or she is unworthy of their confidences, the investigator may as well pack and leave. The reestablishment of a breached trust is most unlikely.

Often, the ancient and traditional beliefs regarding the usage and curative properties of local plants are suspect by the upwardly mobile and "progressive" members of a practitioner's own family. Most of the world's more primitive societies are currently in flux and conflicting value systems are in opposition. In this situation, practitioners of the old ways may guard their knowledge by silence. In parts of Mexico and South America, where inhabitants have become wary due to the influx of Americans seeking drugs, local introductions and the establishment of oneself as a legitimate investigator are essential if any information is to be forthcoming from the community. Also, in semi-urban areas the economic enticement of easy money obtained from the sale of ethnobotanicals may create a situation in which false or purposely misleading data and materials are exchanged for cash.

Considerations in Field Work   Field study, particularly for the purpose of obtaining ethnographic and ethnobotanical data, is an art as much as a science. Social talents and the general acceptance of an investigator determine the quantity and quality of information gathered as much as the investigator's persistence and thoroughness. Successful handling of social encounters, mediating of social roles, and constant credibility are essential. Dealing with social encounters in a way unacceptable to potential informants could mean that access to individuals who might have been of great value will be cut off. A field worker would not intentionally alienate anyone, but sometimes mistakes are inevitable. These can be reduced by intensive study prior to entering the field, and a degree of reservation until one is better acquainted with local customs.

Choosing Informants   Successful field study demands the cultivation of a rapport with those individuals able to provide the most relevant information. Initial introductions within a community are likely to be to expatriate residents and the locally wealthy or powerful. Often, people in these categories do not have the information desired or the social contacts necessary for finding informants. Association with the locally powerful may make the investigator suspect to those who do have the materials and the knowledge being sought. We have found this particularly true in cultures where there are marked class distinctions with minimal interaction between the strata. If the ethnobotanical study is to be realized, the temptation to cultivate friends among the society's elite should be foregone in favor of a polite distance. This manner may facilitate acquaintance with more helpful individuals.

Ethics in Field Work   To what extent should the field worker interfere with the local structure? How much does the investigators' presence interfere unavoidably? In most instances no ethical researcher, having solicited information, would betray confidences from informers when the informers' practices are in conflict with local laws. However, situations may arise when the field worker has to resolve a dilemma. For example, a seriously ill patient under the treatment of a shaman might be cured, or at least have a much better chance of survival, if an injection or some other modern medical aid were provided. Few individuals would hesitate to save someone's life were the means available. The best way to resolve such a conflict is to appeal to the shaman's pragmatic nature. If the patient can be cured and the shaman can gain the credit, then all needs are served. A situation more subtle but insidious occurs when infants are deprived of necessary nutrients by the nature of local, culturally-accepted diets. Contradicting traditional ways will most probably culminate in a rupture of trust and the severing of lines of communication. Criticism of native customs, however well intended, is better left for persons in other roles—at least until necessary information has been procured.

Sex and the Collection of Ethnobotanical Information Few cultures are acquainted with the rhetoric of sexual equality. In many cultures the possession and transmission of shamanistic and related ethnobotanical information may be privy to a particular sex. It may be considered taboo to provide this knowledge to a member of the opposite sex. If, as a potential investigator, you are of the "wrong" sex in terms of the information desired, you would do better to leave the proposed work to someone else. The probability of obtaining accurate data is low.

Clothing and Habits To the inhabitants of ethnically distinct cultures, the sudden appearance of an outsider who attempts to assume a native role in dress and habits can be disturbing, confusing, or irritating. The appearance of an information gatherer can unquestionably affect the information gathered. A socially aware informant is a good source for gaining a social perspective. In some instances, "going native" is considered a form of condescension, or even an insult. In the Guatemalan highlands, where the authors spent some time researching for this book, each Indian village is represented by a particular local costume. The costume serves as a bit of social information—identifying the wearer as a member of a village unit. It may be interpreted as rude and somewhat presumptuous for an outsider to wear ethnic clothing. Dress that connotes membership in a hated or feared social group (e.g., government officials' pith helmets or other such symbolic items) should be avoided. Also, dressing in a manner representative of a higher status than the social group one plans to study is counterproductive.

As well as the problem of integrating oneself into a foreign culture, there is the consideration of language differences. Even if a field worker is reasonably fluent, difficulties in cross-cultural verbal communication will appear from time to time—colloquial expressions, double meanings, and other intricacies of language may be a problem for the investigator. Or, what if, after months of collecting data, an investigator

Above *Market day in Sololá, Guatemala. (Photo courtesy of Dr. Hugh Bollinger)*

Left *Customer in an Indian market in Chiapas, Mexico. (Photo courtesy of Dr. Hugh Bollinger)*

discovers that an informant has been lying? The reasons for this are varied, but the result will be wasted time, as the "information" obtained will be useless. There is always the possibility of malicious intent, if a worker has offended an informant. Or the misinformation may have been due to an informant's lack of appreciation for the researcher's serious effort; that is, to the informer the investigation may have been a joke. Sometimes a distorted sense of kindness on the part of the informant is responsible. The informant decides the visitor wants to hear certain things (any investigator can inadvertently convey this impression). To avoid disappointing his "guest," the informant will supply a plethora of data that is received enthusiastically, but which bears no relation to fact. Sometimes an informer knows less than he implied, and conceals his ignorance by supplying false data. For all of these reasons, it is wise to rely on at least two persons for information, and to compare as a test of reliability.

If the use of a translator is unavoidable, select the person with caution, and preferably on the basis of a recommendation. Unless a translator is perfectly fluent in both languages, communications will suffer. As meanings are unintentionally altered, the quality of the information is affected. Since the researcher will be dependent on the translator, there may be no means for double-checking accuracy. Another problem with employing a translator, especially a local individual, occurs when the subject matter is related to shamanism. Reservations about seeking certain taboo information may overcome any sense of duty to the field worker. One of the authors discovered in Surinam that his interpreter was pretending to ask questions of the Djuka on magic and related subjects, and then providing false information, because he was fearful of becoming a victim of voodoo if he tried to pry into these areas. From his point of view, the less said about it, the better.

Botanical Considerations   Aside from the cultural factors associated with ethnobotanical research, there are some particular botanical considerations. In order to become familiar with the ethnobotany of an area, the field worker should spend at least a year there. Often, botanicals are used seasonally, and a practitioner's memory may not be encyclopedic. Even if it were, pure description is unlikely to be as informative as observation.

Collecting specimens for scientific verification and for the substantiation of scientific names is extremely important. Local climatic and other conditions may make the drying and storage of plant materials difficult. If possible, samples should be catalogued, carefully packaged, and mailed to a holding point. Care should be taken to avoid sending any illegal substances.

Before initiating any field work, an investigator should have a clear idea of the nature of the anticipated ethnobotanical study. Plants serve as food, clothing, shelter, as a basis for shamanistic practices, and in other ways. If the particular area of study to be explored has not been defined before one's arrival in a society, time will be wasted, and establishing one's purpose may be more difficult. Interesting ethnobotanies can be done on cultural uses associated with local technology, as well as on plants with medicinal and magical uses. Unquestionably, careful assessment of an existing situation may affect plans. If a predetermined study appears impossible or unfeasible, then another area might be selected.

Validating the geographical areas in which culturally-used plants grow is necessary. If this is not done, a plant may be assumed to be local when, in fact, it is an article of commerce from a nearby or remote region. This is especially true of materials found in markets; part of a seller's goods may be collected locally, while others are procured through trade.

Probing into ethnobotanical areas of knowledge can be a delicate pursuit, demanding tact and knowledge. The use of plants and their powers—real and assumed—tends to be associated with religious and supernatural capacities. Those having this knowledge are usually esteemed, feared, or both. Persons with such knowledge customarily retain it within a closed kinship system of transmission. Ethnobotany is a relatively open field of study and has been far less thoroughly investigated than other aspects of human societies. Unfortunately, much of the myth and fact relative to plant use in primitive cultures will be lost before thorough investigation is possible. Such cultures are rapidly coming of age in the modern world, or moving toward the threshold of extinction.

Suggested Readings

Adams, R. N., and Preiss, J. J., editors. *Human Organizational Research*. Homewood, Illinois: The Dorsey Press, Inc., 1960. Published for the Society for Applied Anthropology.

Brislin, Richard W.; Lonner, Walter J.; and Thorndike, Robert M. *Cross-Cultural Research Methods*. New York: John Wiley and Sons, 1973.

Pelto, Pertti J. *Anthropological Research: The Structure of Inquiry*. New York: Harper and Row, 1970.

CHAPTER

14

PEASANT MARKETS

In most developing nations the "peasant market" is an integral part of life. In urban and semi-urban areas, marketing is an everyday affair. Sellers have their own stalls, or perhaps a place on the sidewalk for poorer or itinerant vendors. A stimulating array of goods is usually presented, particularly in tropical countries where the odors from mounds of a variety of fruits mingle with those of warm meat, fish, and poultry.

Where there are ethnically distinct groups, such as in the Guatemalan highlands, there may be a major market day just once or twice a week. In Sololá, overlooking Lake Atitlán from a high promontory, Thursday is the main day for marketing, with a lesser market held on Tuesday. Indians come from miles around, usually in scattered village groups. Old men carry stacks of large pottery *ollas* (urns), or firewood, and women come bearing flowers, eggs, live chickens, onions, potatoes, and a respectable variety of other produce. The throng of buyers and sellers is often so dense that one can scarcely move through the crowd. The event is social as well as commercial—providing more flavor to a somewhat rigorous and often tenuous existence.

Various Roles In such markets from Africa to India to South and Central America, most of the traders are women. In many rural areas, market days are staggered, to allow people to utilize optimally the dynamics of the market system.

Speculators. Several different roles can be identified in the typical peasant market. Speculators buy commodities from the women who are either on their way to market or at the farming site itself. Most of these goods are destined for international trade and never reach the peasant marketplace. The woman, who typically acts as the trader for her household, must always try to figure out whether she will receive a higher price by selling to the speculator, or whether she would do better to carry her goods to the market. Many vendors simply sell what they have brought to market and avoid involvement in the larger dealings of the market. After all of their merchandise has sold, they buy what they need for their families and go home. But if a woman is resourceful, she may parlay a small sum into an amount sufficient to secure the status of *reseller*.

Resellers. The resellers move from market to market, dealing in relatively small amounts of any given item; but in total they sell substantial quantities of produce. The potential profit in this technique lies in price differentials between regional markets. The astute vendor increases the value of her or his products by transporting, processing, storing, and bulking of goods. The general native rationale for the function of resellers is that, were it not for these vendors, consumers would have to buy in bulk to hedge against price fluctuations. Essentially, resellers act to stabilize the supply and demand, and thus the market prices. Usually, they buy basic commodities in the back country. There, local markets are not integrated with the national economy; as a result, prices are usually lower. Then they resell the goods from local markets in cities or in larger regional markets.

Truckers.   Truckers have a high status in marketplaces and in their communities. Truckers carry goods and buyers to market. In general, there is not much demand in peasant markets for the bulk hauling of goods. Much of a trucker's income may be derived from transporting villagers and whatever they have for sale. His economic interests are best served by the resellers who regularly travel from one marketplace to another.

Marketplace Dynamics   The dynamics of a market are complex, and not immediately obvious to an outside observer. Generally, markets begin early in the morning, and most transactions have been completed by noon. Even so, in the emptiness of an afternoon market there are characteristically a few hangers-on. Permanent stalls may still be open, and a few old men or women will still be hovering over what remains of their avocados, bananas, or other produce. All activity seldom ceases before dusk, when the last of the buyers and sellers drift away.

Markets typically have an elaborate underlying order and convey a great deal about the society, for the market is the central economic institution of these societies. Economically, markets may be separated into three categories: those where only barter and exchange occur, there being no monetary system; those using money, with prices reflecting isolation from international trade; those having prices influenced by the fluctuations of international trade. Peasant markets in the first two categories are still found in Mexico, Central America, South America, the West Indies, Indonesia, parts of Africa, Malaysia, India, Borneo, and some other countries. But few markets exist solely on a barter economy. The use of some currency, however limited in extent, has become familiar in all but the most isolated areas

Human beings tend to form a commercial network of markets. Often, produce bought in one market will be bulked and sold in another. Regular sellers of staples and hardware are

A peasant market in Togo, Africa. Okra and dried fish are the products being sold. (FAO ph

present everyday, renting stalls on a continuing basis. Produce sellers either rent stalls in this manner, buying from resellers, or they rent only by the day or week. Perishable goods are subject to the greatest price fluctuations. Cloth and hardware are more or less stable in price, although even these items are affected by inflation. Tomatoes, for example, may sell for 8 cents a pound one week, and 25 cents a pound the next if heavy rains cause a loss of crops.

Characteristically, sellers of a given type of goods are found clustered together. This permits a quick establishment of prices for the day. The practice also enables sellers to see all of the potential customers, to service the regular customers, and it makes possible the floating of prices throughout the day—with a quick consensus on increases and decreases.

A seller will usually have a quite regular trade with the same individuals, and gives concessions to these customers to secure their continuing business. Such concessions usually consist of special prices, credit, and particularly quality. The seller's cooperation in these respects is meant to guarantee that she or he will have a faithful clientele when the market is glutted and when prices are generally low. If there has been a bumper crop of potatoes, for example, a group of dependable customers will mean more sales than if the surplus had to be sold in even competition.

A successful rural market will attract retailers from the towns. They bring stocks of their shoes, cloth, and such manufactured items. Considerable amounts of goods may be sold to rural inhabitants whose visits into town are infrequent. Clothing, shoes, and cooking utensils are especially popular. Hustlers are also common, selling worthless "remedies" for worms, snakebite, and numerous other ailments. Usually, they are less colorful versions of "medicine men" who once circulated in the United States. They claim efficacy for their potions in treating almost anything.

Markets in peasant economies are relatively heavily taxed. A seller must rent a stand in the marketplace, which is usually municipally owned, and a local seller's tax must be paid. Traders are also taxed for taking livestock to market, for butchering at the market, and for selling alcoholic beverages and locally grown tobacco.

Peasant markets are difficult to incorporate into the national economy, a characteristic having social and economic implications. Buyers and sellers in rural markets are seldom aware of international trade, and generally do not produce goods involved in it, except for those items sold to a speculator.

A market fulfills numerous purposes besides being a place

for exchange, buying, and selling. It provides a common ground for gossip among those selling their wares and produce. It broadens a young person's possibilities for courtship, by providing a place for interactions. When the inhabitants of remote villages come in to market, they have an opportunity to go to church, to visit shops selling items unavailable where they live, to visit public health clinics, and to register births and deaths. The main purpose of a market is to provide a place for commerce; yet, in reality, the peasant market is a multi-faceted social institution.

Currently, peasant markets exist primarily in developing countries, and are associated in our minds with tropical areas —a more or less correct assumption today. However, similar markets were once found throughout Europe and in other areas where sedentary lifestyles promoted the commerce of necessities and surplus goods. In North America, open markets used to offer a variety of meats, fruits, and vegetables. In character these markets were probably reminiscent in many ways of existing peasant markets. *The Market Book* offers an insight into the foods available a hundred years ago in a New York market.

> Cucumbers are abundant. Calabashes, or gourds, also grow there: they are half as long as the pumpkin, but have within very little pulp, and are sought chiefly on account of the shell, which is hard and durable, and is used to hold seeds, spices, etc.
>
> Turnips, also, are as good and fine as any sandrapes that are raised in the Netherlands. Of beans there are several kinds. The Turkish beans which our people have introduced there grow wonderfully; they fill out remarkably well, and are much cultivated. (De Voe, 1862: p. 24).

Most peasant markets of today have changed little over the past few decades in terms of dynamics, but there are observable differences in the goods and in the physical character of the markets. For example, most Mexican markets have evolved from open arenas, with wooden stalls and many vendors seated on the ground, into enclosed, cement buildings with permanent cement stalls and often a background accompaniment of piped-in music. A plethora of manufactured goods has become available, a majority of which are inexpensive plastic or tin items meant to appeal to unsophisticated shoppers with limited budgets. There is customarily an abundance of religious merchandise, from plastic saints to framed prints of religious figures. Other goods include plastic shoes, cheap jewelry, tin kitchen utensils, and a variety of miscellanea. Produce and meat are predominantly unchanged in the way they are handled, except that the foods sold in permanent stalls are often less open to price bargaining.

A Representative Market Economy   The market economy of South Dahomey in Africa may be considered as a representative example. (Tardits et al. 1962). Although the political structures in many parts of Africa have changed greatly within a relatively brief time period, it is safe to assume that the economic realities of life at the rural level have been considerably less affected—including the nature of peasant marketing. In South Dahomey, markets are numerous, and it is difficult "to walk five or six miles" without encountering one. The markets are almost entirely a female domain; the males in the rural population are engaged in farming or fishing. In the society of South Dahomey, a man is expected to grow crops and produce food from his land, which is customarily of patrilineal heritage. The men are seldom seen in a marketplace except as buyers.

In contrast to the typical male, the female is a relatively mobile member of the society. When she marries, she moves into her husband's dwelling. During the first few years of marriage, she typically accepts most of the household chores, such as carrying water, washing, and mending her husband's clothes, cooking meals, and managing all other aspects of the household. However, after the birth of a child or the introduction of another wife, a considerably lesser household commitment is demanded of her. She has time to pursue outside activities, the most desirable of which is marketing. The profits she earns are her own. She may buy goods from her husband, and the arrangement remains businesslike in most cases.

Women who are shrewd in business can fare quite well in the marketplace. Corn is a leading speculative commodity. Those who are aware of market trends and able, financially, to do so will buy corn in March or April, when it is plentiful. They then resell it later during June and July, just prior to the next harvest, when prices are highest. Generally, most of the women are successful capitalist entrepreneurs. If they receive word that a ship bringing supplies will be delayed, they may buy up stocks of cigarettes, sugar, and other valued items, thereby controlling the prices, which may double or triple. Women who can afford the initial expense may also become wholesalers, dealing primarily in staple crops, such as corn, cassava, yams, groundnuts, beans, and palm products. However, there is a far greater number of retailers who sell various

A sugar loaf stall on market day in the Oaxaca valley of southern Mexico. (FAO photo)

commodities obtained in the bush as well as crop foods. Some typical items from the bush are wild fruits, packing leaves, medicinal plants, and wood. Dealing in these gathered items is especially characteristic of vendors lacking capital for more lucrative enterprises. Older women are more likely to sell herbs, perhaps because it is assumed that young women would not have acquired the knowledge necessary for securing effective herbs.

In any Dahomean, and generally in almost any peasant marketplace, the odors of cooked foods mingle with those of meats and fresh produce. Husbands whose wives are engaged in marketing will frequently eat meals at the marketplace, where they choose from such fare as grilled peanuts, smoked fish, boiled *akasa* (a corn paste), *abobo* (beans mashed and mixed with pimento and corn oil), *acra* (mashed sweet potatoes), *doko* and *atakre* (beans and sesame croquettes), and a variety of stewed dishes. The women selling cooked foods comprise the youngest group in the market, and the most numerous. Eventually, almost all of them will sell *akasa*, as it is a perferred staple and any girl learns to make it while she is still a small child.

Although market sites are the heart of commerce, marketing is not limited to the marketplace. Women selling their goods are encountered along the sides of country and major roads, at the junctions of pathways, and in other places where potential customers are anticipated. Reasons for this are various. It may be that a woman prefers to remain closer to home. Or she may be in mourning and therefore prohibited from entering the marketplace until the final death rituals have been completed. Some women, when the busy morning session in the marketplace is over, will gather up what remains of their produce and attempt to sell it door to door. This is also true in Latin America, where Indian women circulate during the slow market hours, carrying their baskets of eggs, avocados, or whatever.

In most peasant markets competition is relatively stiff, especially among vendors selling the most abundant products. Buyers will usually be influenced by appearance, and so each woman labors to convince the potential customer that her vegetables are fresher, that her *akasa* tastes the best, or that her avocados are the largest. Bargaining is customary in peasant markets—providing the seller with a predominantly enjoyable interaction, even when insults are included, as long as these remain within acceptable limits. The seller rarely loses, unless she has the misfortune to have stocked a surplus item. Prices are highest during the early morning, and taper off after the height of activity, but then rise again during the quiet of the afternoon. In the market in Panajachel, Guatemala there are always a few older women still selling produce, fruits, and eggs until dusk. During the quiet interlude before the morning frenzy, prices are inevitably higher, as the competition is so markedly reduced.

Bargaining

Bargaining is at least as much a social as an economic tradition. A buyer should know what the going cost for an item is for the day, or he or she will be at a disadvantage in arriving at a fair price. If, for example, avocados are selling for 5 or 6 cents, the buyer may offer 3 cents. The seller will then respond: "What? Oh no, these avocados are worth at least nine. Look at them, they are so large, and *very* tasty." The buyer will respond by deprecating the produce, and perhaps the vendor's family ancestry as well, if the haggling continues. Usually a price satisfactory to both parties is agreed upon, or the customer will try his or her luck with another vendor.

Sellers in peasant markets are clever at convincing customers that they have been given a special bargain. As in supermarketing techniques, packaging can make a difference. In Guatemala, Indian women sell raspberries and strawberries in containers made of banana leaves which are fashioned to appear more commodious than they are. Shaped like an angular funnel, the small leaf basket opening exhibits the best and the majority of fruits. The few fruits underneath are usually smaller, unripe, or both. Another mechanism in the seller's favor is that beans, fruits, and other foods sold by weight are placed in primitive balance scales consisting of two baskets or tin pans suspended on either end of a stick. A measured weight is used to balance the quantity of food desired. However, the baskets are seldom suspended evenly, and of course the seller will put her goods into the one that hangs lowest.

In an analysis of marketing systems, Mintz (1959) suggests that the existence of markets is a sign of nonself-sufficiency. This is, of course, true to an extent; but the development of markets is a complex process arising not only from the need to exchange goods, but to fulfill certain social functions as well. Also, the concept of pure self-sufficiency, in the sense of a cultural group's isolation from others and an absence of dependency regarding the exchange of goods and services, is practically nonexistent today. A few primitive, vanishing societies are the rare exceptions.

Functions of the Market A market not only has a social role within the context of the culture in which it evolved; it is an effective locus for the introduction of innovations or extracultural material objects and even ideas. As a place where different villages, ethnic units, and social strata may come together, the market-place can aid in the proliferation of foods and merchandise not integral to the culture. For instance, if a nutritionist were to convince only a few farmers to grow an unaccustomed but beneficial crop, the place of its dispersal would presumably be the market. Whether or not the food gained popularity, it would be given exposure.

Also, imported manufactured goods are frequently introduced into the rural community via the market place. In most rural societies manufactured goods are esteemed, and one need only make them available. In Surinam, Djuka women keep vegetables, fruits, and grains in plastic tubs stored in the recesses of their simple thatched huts; almost anything manufactured is prized.

Local markets do reflect the conservatism of traditional values. For instance, a Latin American market without beans and corn would be rare. Even if equally nutritious foods were substituted for traditional ones (wheat and oats in place of corn, for example) the acceptance of the different items would be grudging. Most staple foods in peasant societies have been integrated in various ways into the cultural fabric; it is not easy to successfully introduce surrogates. Dietarily, a Guatemalan Indian might fare better on soybeans and wheat than on corn and beans. But attempts to change culturally established eating habits usually meet with total or partial failure. As a central artery of social interactions, peasant markets reflect innovation. As upholders of the status quo, especially in matters of food, the marketplace is a source of quick reference on the accustomed meats, fruits, vegetables, and grains of an area.

From Peasant Market to Supermarket   In the United States, in much of Europe, and in other "developed" areas of the world, there are no peasant markets. Food distribution passed through a stage wherein the major outlet was through small food stores, often owned or managed by a single family. In less affluent countries, it is still possible to observe this first step away from the peasant market. In Paramaribo, Surinam, for example, there is a large central market, with stalls and piles of fruits and vegetables; but there are also dozens of small food shops selling minimal amounts of produce, canned goods, and sundry commodities.

During the early 1900's, chain food stores were started in the United States. These became popular, and threatened the livelihood of the small-store owner. Between 1914 and 1930, corporate food chains grew from 500 parent companies with 8000 units in 1914 to 995 parent companies with 62,725 retail

Scene in a typical market in Bogota. (FAO photo)

units in 1930 (Zimmerman, 1931). During this time, there was considerable antagonism toward the chains. Local business owners considered them a threat to private industry. From 1925 to 1932, more than 300 chain store tax bills were introduced in legislatures.

Then, in the early 1930's, the first two "supermarkets" were opened in New York: King Kullen and Big Bear. The term "super market" was not in use at the time, and seems to have evolved with the industry—coming into vogue shortly after the appearance of these two major stores. By 1934 the term was already being used. Today, a supermarket is a departmentalized retail business specializing in food. The store usually has ample parking space, a notably large gross income, and a self-service policy.

Supermarkets are now an accepted part of the American lifestyle, and most Americans purchase their food from them. Due to the great quantity of goods sold, supermarkets are able to profitably offer foods at prices lower than those of smaller chain or privately-owned stores. But some consumer problems have grown with the proliferation of supermarkets. Prepackaged foods are increasingly popular, yet the nutritional quality of these foods is not always adequate. Such foods are often marketed on the basis of their attractive appearance, sweetness, or various artificial flavorings. Frequently, the protein and vitamin content is low.

Suggested Readings

Benedict, Peter (1972). "Itinerant Marketing: An Alternative Strategy," in *Social Exchange And Interaction*. (Museum of Anthropology, University of Michigan: Anthropological Paper No. 46). Ann Arbor: The University of Michigan.

Charvat, Frank J. *Supermarketing*. New York: The MacMillan Company, 1961.

De Voe, Thomas. *The Market Book*. New York: Burt Franklin. (This book was originally published in 1862).

Ellis, Effie O. "The Family: Nutrition and Consumer Problems," in *U. S. Nutrition Policies in the Seventies*. San Francisco: W. H. Freeman and Company, 1973.

Hodder, B. W. and Ukwa, U. I. *Markets In West Africa*. West Africa: Ibadan University Press, 1969.

McEwen, Robert J. "Consumer Problems in Relation to the Food Industry," in *U. S. Nutrition Policies in the Seventies*. San Francisco: W. H. Freeman and Company, 1973.

Mintz, S. "Internal Market Systems As Mechanisms of Social Articulation," in *Intermediate Societies, Social Mobility and Communication*. Seattle: American Ethnological Society, 1959.

Sudarkasa, Niara. *Where Women Work: A Study of Yoruba Women in the Marketplace and in the Home.* (Museum of Anthropology, University of Michigan: Anthropological Paper No. 53). Ann Arbor: The University of Michigan, 1973.

Tardits, Claudine, and Tardits, Claude. "Traditional Market Economy in the South Dahomey," in *Markets in Africa.* Edited by Paul Bohannan and George Dalton. Illinois: Northwestern University Press, 1962.

Zimmerman, M. J. *The Super Market.* New York: McGraw-Hill Book Company, Inc., 1955.

CHAPTER

15

PSYCHEDELICS, SHAMANISM, AND HEALING

One of the dilemmas of human consciousness is how to achieve stability on the one hand while being attracted to novelty on the other. Individual and cultural perceptions of the world vary, but the way human beings perceive various experiences is also defined and restricted by biological limits. When human biochemistry is altered, the nature of human perception is also altered. In our culture, alcohol is the primary substance for accomplishing this. However, many others have been used in the past and are still in use today. Different drugs produce different effects which are qualitatively predictable, to some degree, for each kind of drug.

Hallucinogens   Hallucinogenic drugs have long been a part of the religious experience in many cultures, especially materially primitive ones. The most important of the hallucinogens are: *mescaline,* found in the peyote (or *peyotl*) cactus (*Lophophora williamsii*); *psilocin* and *psilocybin,* present in several mushrooms, the most commonly used are *Psilocybe mexicana* and *Stropharia cubensis;* and *LSD* (lysergic acid diethylamide), which is easily synthesized in the laboratory, and which may be derived from the drug ergot, obtained from a fungus (*Claviceps purpurea*). *Banisteriopsis caapi,* a tropical South American vine, yields *harmine* and *harmaline.* Indians boil the vine in water to produce a psychedelic drink. *Opium,* from opium poppies (*Papaver somniferum*) causes a euphoric experience. Marijuana (*Cannabis sativa*) yields a drug, *tetrahydrocannabinol,* that is mildly psychedelic in effect.

In recent years, particularly during the 1960's, psychedelics gained popularity in Western cultures, and debates raged over whether or not the drugs cause psychological or physiological damage to users. In some primitive societies, psychedelics are employed in treating the mentally ill, and have been the subject of research in this area among modern investigators. However, even in the most primitive groups, care is taken not to administer the drugs to psychotic persons. The arguments supporting the use of psychedelics are based on the value of exploring varying perceptions of reality. Recipients of such drugs have served as models for psychiatric studies, because there are similarities between drug-induced and naturally-occurring psychoses.

There is a strong bias against such drugs in Western societies, possibly due (as suggested by Michael Harner, 1973) to the witchcraft practises of the Middle Ages in Europe and the force of eradication of these practises by the Inquisition and Witch Trials, both in the New and Old Worlds. Practitioners of European witchcraft commonly used nightshade

(*Atropa belladonna*), henbane (*Hyoscyamus spp.*), mandrake (*Mandragora spp.*) and locoweed (*Datura spp.*). These plants are all members of the *Solanaceae*, and contain the psychedelic substances *atropine, hyoscyamine*, and *scopolamine*—highly toxic alkaloids that cause deep sleep, paralysis, or death from overdose. In the New World, the Inquisition prohibited the use of plants containing *mescaline* and *psilocybin*. Consequently, Western societies in the Americas had little cultural experience with psychedelics until recent times.

During the latter part of the 19th century, psychedelics began to attract attention once again. Aldous Huxley, in *The Doors of Perception*, and others began investigating the drugs' potential in the exploration of human consciousness. In 1896, the active principle in mescaline was isolated. In 1919 it was recognized that the molecular structure of mescaline is similar to that of the human neural hormone *epinephrine*, a discovery that marked the turning point in serious inquiry. Substances that chemically transmit impulses across neural synapses include *acetylcholine, epinephrine, norepinephrine*, and *serotonin*. The primary point of psychiatric inquiry was whether or not the active elements of psychedelics cause the same effects as these intrinsic substances; and if so, are psychoses a result of errors in the body's quantity of metabolic output?

The hypothesis that psychedelics were mimicking the effects of internally-produced substances was strengthened when it was found that breakdown products of epinephrine could be isolated from the urine of subjects who had taken LSD. This fact was known indirectly by primitive peoples for some time, as they frequently drink their own urine to enhance the effects of *Amanita muscaria* (a mushroom) and *Banisteriopsis*. It is known that depressive and catatonic states are caused by a deficiency of serotonin, and that excessive quantities of serotonin result in hallucinations. However, the hypothesis remains unproven.

The Psychedelic Experience Generally, there are three kinds of variables in the psychedelic experience: the set and setting during drug use; the properties and potency of the drug; and the personality of the individual taking the drug. Different drugs tend to be characterized by their own sets of properties. The *indole* group (found in LSD, psilocybin, mescaline, and serotonin) changes self-perception of the body and creates the illusion of "melting" and dissociation. Auditory hallucinations are common; the sense of time becomes distorted; and the distinction between the subject and the physical surroundings is

Left Psilocybe mexicana. *The most widely used psychotropic mushrooms of the Mazatec Indians of Oaxaca, Mexico.*

Below *Peyote (*Lophophora williamsii*).*

lost. The *atropine* group (found in belladonna and loco-weed) produces visions in brilliant colors, often a sensation of flight, a sense of clairvoyant perception, time distortion, and loss of the ability to distinguish reality from fantasy. The DMT (dimethyl tryptamine) group (*Banisteriopsis* and *Psychotria*) are similar to the above.

Physical symptoms include dilation of the pupils, peripheral vascular constriction, fever, a rise in systolic blood pressure, the nonspecific arousal of brain waves, and insomnia after the experience. None of the true psychedelics are physically addictive. The duration of a "trip," as the psychedelic experience is often termed, may vary from several to eight or more hours—depending on the particular drug and the quantity ingested.

Drug Use in Primitive Tribes   In very few areas of the world is the use of hallucinogenic drugs still practised under aboriginal conditions. One of the last of these is among tribal groups in the jungles of the upper Amazon Basin in South America. Shamanistic rituals in this region typically involve the drinking of a tea brewed from *Banisteriopsis* vines. The drink is known as *yaǵe* or *yaje* in Colombia, *caapi* in Brazil, and *ayahuasca*, meaning "vine of the dead" in Quechua, the language spoken by some Indian tribes in Ecuador and Peru. Although species of *Banisteriopsis* grow in Central America, Mexico, and the southeastern United States, the plants have evidently not been used as psychedelics in these areas.

All *Banisteriopsis* grow wild as vines which climb over trees for support. Some tribes cultivate the vines in gardens as well. *Psychotria viridis,* also containing a powerful hallucinogen, is added to the drink in some areas, especially in eastern Ecuador and eastern Peru. The biochemistry of the drink made from *Banisteriopsis* is poorly known, partly because another plant—either *Psychotria* or an unknown—is almost invariably added. *Harmaline,* the active hallucinogenic substance in *Banisteriopsis,* has been recognized for some time; but other alkaloids are also present, such as N, N-dimethyltriptamine, found in *Psychotria* and some *Banisteriopsis.*

Banisteriopsis "tea" is taken by shamans to enable them to commune with spirits and other aspects of the supernatural, and to influence certain supernatural entities, such as those attacking a diseased person. Non-shamans are also allowed to take the drug, for the purpose of gaining supernatural powers, engendering visions, or to accompany a shaman in a ritual performance. In some cultural groups, as with the Peruvian Cashinahua, the drug may be taken in groups. Others, such as

Banisteriopsis caapi, *one of the
two plants commonly used in
preparing the brew known as*
yagé, ayahuasca, caapi, *or
among the* Jivaro, naternä.

the more solitary Jivaro, prefer to take it individually.

Among any of the social groups which accept the use of hallucinogenics, unless there are rigid restrictions of drug use, almost anyone can achieve a trance state considered requisite for the practise of shamanistic rites. In Jivaro society, the availability of *Banisteriopsis* accounts for the relatively high percentage of shamans. Because of the depth of their psychedelic experiences, the Jivaro believe that ordinary perceptions of the world are false, and that only under the influence of an hallucinogen does reality manifest itself! This philosophy is reminiscent of that expressed by Don Juan in Carlos Castaneda's books (see Suggested Readings), except that the Jivaro are willing to accept only one reality, that of the supernatural, instead of allowing for many.

Kenneth Kensinger, in describing the experience of drinking *Banisteriopsis*, or *nixi pae*, among the Cashinahua Indians of Peru, recounts an informant's experience (Kensinger, in Harner, 1973).

> "We drank *nixi pae*. Before starting to chant, we talked a bit. The brew began to move me and I drank some more. Soon I began to shake all over. The earth shook. The wind blew and the trees swayed. . . . The *nixi pae* people began to appear. They had bows and arrows and wanted to shoot me. I was afraid but they told me the bows and arrows would not kill me, only make me more drunk. . . . Snakes, large brightly colored snakes, were crawling on the ground. They began to crawl all over me. One large female snake tried to swallow me, but since I was chanting, she couldn't succeed. . . ."

Among the Cashinahua, any male is permitted to drink the brew of *Banisteriopsis*, also known as *ayahuasca*. Its use is optional, and there are some who never drink it, whereas others will do so whenever an opportunity arises. Usually the preparation of *ayahuasca* is left to one or two men in a village, although most male members of the society know how to prepare the drink. When the decision is made to imbibe it, an entirely arbitrary choice made by general consensus, the "host" goes into the jungle, where he cuts about six feet of *Banisteriopsis* and a few branches of *Psychotria*. The former is sectioned into 6- to 8-inch pieces, pounded lightly, and placed into a cooking pot with a capacity of two to four gallons; and then the leaves and buds of *Psychotria* are added. This mixture is simmered for an hour or so, then ladeled into smaller containers for cooling. When it is time to partake of the drink, each man chants over it, rapidly drinks about a pint, then joins the others to wait for the effect to begin. If he wishes a "good trip," as the Cashinahuas say, he may drink a second pint.

Few of the Cashinahua questioned about their experiences with *ayahuasca* described them as pleasant. If that is so, why do the men repeat their use of the drug? Supposedly, it reveals to them past and future events, or information about persons who are not present. Hallucinations are considered to be the experiences of one's "dream spirit" (one of five spirits that the Cashinahua believe each person to have), and may warn of future incidents, as well as recalling those of the past.

Among South American Indians, whether Cashinahua, Jivaro, or others, the psychedelic experience is essentially a means of communicating through channels not open to ordinary senses. The alteration of perception, although it may in itself be a fearful experience, is considered a means for divining things which cannot be otherwise understood. Even if the experiences undergone while "tripping" are not "real," the illusion of reality is so marked as to seem but one aspect of a total reality, or, as with the Jivaro, to supersede the reality of the everyday senses. In an environment which is, in most respects, beyond the control of technologically primitive human inhabitants, any means through which an advantage or a presumed advantage is conferred in the goal of survival is likely to be adopted. Whether or not taking psychedelic drugs actually *is* an advantage, the Indians believe that the practice provides an aid in their struggles against disease, famine, enemy attack, and the other exigencies with which they must cope.

The use of *ayahuasca* is not limited to ethnically distinct tribal groups. In areas of South America where some tribal peoples have migrated into cities (and thence to slums), cultural beliefs have remained. (de Rios, in Harner, 1973). In the Amazon city Iquitos, in Brazil, *ayahuasca* was used to cure diseases, especially those of a "magical," or psychosomatic origin. The shaman, or *ayahuasguero*, is expected to reveal the cause of the patient's illness through contact with him or her during the drug experience. An illness that seems to be excluded from the healing effects of *Banisteriopsis* drinks both in jungle tribes and in cities is psychosis.

In the Sierra Mayateca, a range in the northeast of the Mexican state of Oaxaca, mystical connotations are associated with a mushroom (*Psilocybe mexicana*). The "mushroom cults" of that region are a blend of paganism and Christianity. Today, the cults are withering, and are restricted to only two or three small villages. In past centuries, during the dominance of both the Mayan and Aztec civilizations, mushroom use was common, especially among the priestly classes.

In the Mayan culture, mushroom use was technically restricted to the priesthood. Among the Aztecs, the mushrooms were a ubiquitous element in the society. It was reported in the *Cronicas* of the Conquest of the New World that mushrooms were passed out to approximately 500 thousand participants in the coronation of Moctezuma. The British anthropologist Gillsen (see Munn, in: Harner, 1973) has associated the wide and relatively indiscriminate use of mushrooms by the Aztecs with their blood cult of sacrifice and the disintegration of social order that was occurring at the time of the Spaniards' arrival.

Mazatec Indians of Oaxaca believe that the mushrooms may be eaten *only* at night, as madness will ensue if they are consumed by day. Ordinarily, the experience is a family affair, with the father officiating, although a shamaness may be involved during curative affairs. Aunts, uncles, and other close kin may be invited to participate. When everyone is present, the father will lock the doors, close all of the windows, and then distribute the mushrooms. Each person ingests two mushrooms, one being male and the other female in representation. The mushrooms are always blessed with tobacco and copal smoke before being eaten. The father then chants, sings, and recounts his visions to the assembled group throughout the night. Other members are allowed to express their visions as well. It is a conducive environment to the psychedelic experience. The darkness fosters hallucinations and, since visibility is poor, voices may seem even more remote and altered than if heard while "tripping" during daylight.

Among the Mazatecs, anyone may indulge in the mystical "journey" of eating mushrooms, but the shamans are customarily the only ones who do so on a repeated basis. The shamans are thus considered experts and guides toward realizing what is sought in the experience. Mushrooms may be taken for curing illness, or may be a means of resolving personal conflict or some other problem. During the experience, the mushrooms are said to "speak" through the human vehicles. That is, the thoughts expressed are not considered those of the speaker, but are supposedly inspired by a higher wisdom released through the person's hallucinogenic "trip."

Mushroom use is associated with problems, whether of mental or physical illness, or ethical dilemmas. Until recent times, mushrooms were a major element in combatting disease. Indians who subscribe to their powers still believe fully that eating mushrooms will aid in the cure of syphillis, cancer, epilepsy, stomach ulcers, and other ailments. The introduction of modern medicine into the areas of mushroom use

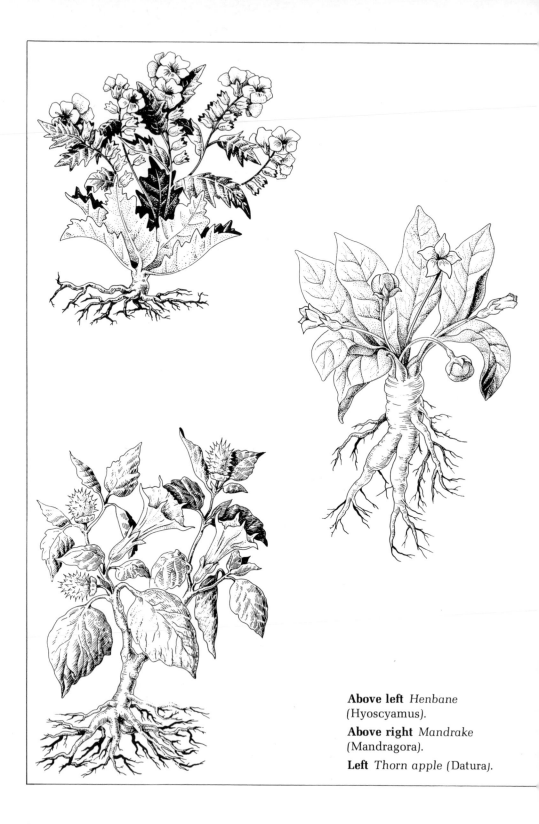

Above left *Henbane*
(Hyoscyamus).

Above right *Mandrake*
(Mandragora).

Left *Thorn apple* (Datura).

must have been an influence in the plant's waning significance. Today, there are also cultural and legal sanctions against the drug.

Hallucinogens and Witchcraft

The social role of hallucinogenic plants has not been limited to non-Western cultures. Hallucinogens were once commonly used in Europe. As previously stated, the drugs' insignificance in Western culture until recent years is probably due, in part, to church prohibition, strengthened by the appalling tactics of the Inquisition. But European witchcraft, by no means an imaginative fiction, was based upon the effects of psychotropic plant substances. Although witchcraft has had a surge in popularity, modern covens are relatively sober and ignore the importance of psychedelic drugs to the earlier practitioners of witchcraft. Yet many of the beliefs popularly associated with witches, such as flying through the air, can be logically connected with the effects of hallucinogens.

Many hallucinogenic plants contain *atropine*, which has the quality of being readily absorbed through the skin. European witches made ointments of various hallucinogenic plants, such as belladonna and henbane, and rubbed them on their skin. The resulting trance was usually characterized by the illusion of flying. During these "trips," the participants imagined themselves to be flying away to commune with spirits and devils at *Sabbats*. The seeming reality of these experiences was such that, even faced with torture or death by the Inquisition, witches would frequently refuse to recant their belief in the experiences. The resulting loss of human life was not small—perhaps in the hundreds of thousands, if one includes those who protested their innocence as well as the ones who clung to the validity of their witchcraft.

The sensation of flying is apparently a common effect of the drugs used by medieval witches. Michael Harner (1973) relates numerous instances of witches who had been observed in their trances, insisting that they had flown to distant places or separate worlds. Later experimenters experienced the same illusion during their dream-filled sleeps. Gustav Schenck (1955) commented on his experience with henbane that "the frightening certainty that my end was near through the dissolution of my body was counterbalanced by an animal joy in flight. I soared where my hallucinations—the clouds, the lowering sky, herds of beasts, falling leaves which were quite unlike any ordinary leaves, billowing streamers of steam and rivers of molten metal—were swirling along."

Not only did hallucinogenics comprise a foundation for the religion of witches, but evidence suggests that the basis for *lycanthropy*, the perception of oneself as a wolf or other predatory animal, or the belief that one can be transformed into that state, was similarly induced. The symptoms of a person under the effects of atropine coincide with those of a Greek admonition on what to look for in a potential were-wolf (Adams, 1844). Also, Hesse (1946) observed that "A characteristic feature of *Solanaceae* psychosis is furthermore that the intoxicated person imagines himself to have been changed into some animal. . . ."

So the inception of two major aspects of European supernatural mythology stems from the use of psychotropic drugs. An interesting point made by Harner is that the nature of the drugs influenced the nature of associated rituals. The Solanaceous plants used in Europe are too powerful to allow ritual functioning during their use, thus the witches had *Sabbats*, the drug experience, and *Esbats*, meetings of a "business" nature. In the Americas, where most of the psychedelics are not incapacitating, the rituals associated with them occur under the influence of the drug itself. It seems quite possible that much of the belief structure of the supernatural in many cultures is related in some way to the use of psychotropic plants.

Healing

The employment of plants other than psychedelics as medicinals has been a facet of almost any pre-modern culture. Some of the plants are probably worthless, except from a psychological standpoint, whereas the curative properties of others have a basis in fact. The plants' healing value must have been determined by experiment and by accident, the knowledge of botanical medicinals accruing from one generation to another. American Indians say they "received this knowledge in dreams" (Densmore, 1974), a not improbable rationale in society in which such knowledge has accumulated gradually, subtly, and often indirectly.

Among American Chippewa Indians, the goals of a long life and good health are traditionally paramount to any others; so, logically, the role of a healer is esteemed. *Manido*, or spirits, are deemed responsible for curing through the medium of a medicine man. One method of treating illness uses no material remedies. The other method, involving carefully guarded secrets, makes use of various plant materials. It is believed that every tree, bush, and any other plant has a potential medicinal use. Densmore found that people of the same tribe had differing names and ascribed different functions to the same plants; the remedies were considered individual. Customarily, one individual would specialize in the treatment of only one or two diseases. It is worth noting that the Indians claim to have had many fewer diseases before white men introduced theirs. For this reason, it was much easier, they say, for a healer to remain so highly specialized.

The Midewiwin, that class of the Chippewa who hold the secrets of plant medicinals, say that they "follow the way of the bear." Bears, because they eat cherries, acorns, and roots, are believed to know the use of plants and are supposedly the leaders of animals in the use of herbal medicine. Thus, if one dreams of a bear, that person is meant to become an expert in plant medicinals.

Medicinal roots may be considered significant because of the appearance alone. Any plants having a divided taproot are especially valued, because the roots resemble the legs of a human. Also, sterility or fertility can be a factor in the functions attributed to a given plant. A treatment for dysentery calls for a flowering mugwort (Artemisia dracunculoides), whereas a remedy for strengthening the hair specifies a sterile plant of the same species.

The nature of the soil is also a factor. Plants should be gathered during the late summer or early fall, when they are completely developed. During these periods, trips may be made for gathering plants growing in various kinds of soils. When a medicinal plant is dug from the earth, a small hole is made in the ground beside it and tobacco is placed within while the procurer speaks to the plant, as a sort of offering to insure the best results from its use.

Roots are the most commonly used parts of plants. Usually, the entire root is valued, but in some cases the properties desired may be restricted to particular sections. In dogbane (Apocynum) the "elbow" of the root is preferred, and sometimes the fine white hair roots are used. When stalks, leaves, or flowers are selected, they are tied in bunches and left hanging to dry. When bark is needed, it is best gathered while the sap is in the tree, whereas roots are collected at a time when the sap is not flowing.

An impressive variety of plants are part of the medicinal lore of the Chippewa. Balsam fir (Abies balsamea) provides a cure for headache. Eye diseases are treated with the roots of Jack-in-the-pulpit (Arisaema triphyllum), Dogwood (Cornus alternifolia), Alder (Alnus incana), and others, while cataracts are specifically countered with Red Raspberries (Rubus strigosus), and styles or inflammations of the eyelid are treated with Squirrel-tail (Hordeum jubatum) and Twisted-Stalk (Streptopus roseus). Dogbane (Apocynum androsaemifolium), Asters (Aster nemoralis), and others are used against earaches. Joint diseases, such as rheumatism, may be treated by using twigs of Red Cedar (Juniperus virginiana) or dried flowers of the Pearly Everlasting (*Anaphalis margaritacea*). The effects of snakebite are supposedly countered with a poultice of the root of a Rattlesnake Fern (*Botrychium virginianum*); and the dried, powdered leaves of Fireweed (*Epilobium angustifolium*) are moistened and applied to burns. (For a complete listing, see Densmore, 1974.)

Healing and Power   Among the Indians of Latin America, herbal knowledge also carries social and financial rewards (Lewis, 1970). There tend to be three categories of medical practitioners, besides licensed doctors. *Curanderas* are usually women who know how to utilize herbs. *Mágicos* are males who fulfill a similar function, but employ spiritualism and magic, thereby commanding more respect and higher fees. A third practitioner, "el doctor," is simply a man without medical training who poses as a doctor, extracting even higher fees than the former two. Curanderas cure a child's ailment known as "fright" by sprinkling powdered cedar, palm, and blessed laurel over the forehead, breast, wrists, hand palms, and nape of the neck, and into the child's nostrils. The *mágicos* (shamans) use the same herbs as the *curanderas;* but due to the mystical element, the mágico is both feared and respected. (This is not to say that a *curandera* cannot invoke the supernatural; a friend of the authors was soundly cursed by a *curandera* in the central market in Oaxaca merely for examining some of the latter's herbal wares).

In the context of healing and other shamanistic practices, plants may provide the basis for considerable social power in primitive societies. This is true primarily because in these areas the layman is contending with difficulties beyond his or her immediate ability to resolve. That is, the *structures* of a nontechnological culture—in order to deal with those elements beyond individual control (see Adams, 1975)—have generated the development of social roles filled by persons purported to exercise some control over these situations, even if, in reality, they do not. From this perspective, it is easy to understand why help is elicited from the same person for basically different problems. The person who is vexed or afflicted is removed from a position of direct responsibility. The effects of hallucinogenic plants are important in this respect, for the plants produce the illusion of temporarily breaking down the structural barriers, of allowing both the shaman and the patient to enter into a direct spiritual communication for the purpose of solving a problem. Even a *curandera* satisfies this requirement to some degree, by providing the pragmatic application of her knowledge toward the solution of a conflict—whether it is ensnaring the object of someone's affections or curing an illness. In all human interactions, the parameters of every individuals' existence vary from those of all others; what comprises restriction for one may not for someone else. Thus, those using plants for curative or mystical purposes are acknowledging the variation and limitations defining human existence, whether internal or external, physical or mental.

Pharmacology   As the use of botanicals for healing developed with prehistoric humans, we can only speculate about the origins of pharmacology. We know more about the diseases of early human beings than we do of the treatments, because of the evidence present in skeletal remains. But it is a reasonable supposition that people began to use plants in their search for means to cope with an often mysterious and hostile world. As we pointed out in our discussion of psychedelics, the inducement of a mystical experience may be an attempt to separate oneself from everyday reality, thereby opening the way to new avenues of healing and problem-solving. When it happened that the ingestion of a plant substance was accompanied by a markedly positive physiological reaction, then that plant would be associated subsequently with healing.

Whereas folk remedies came into disrepute with heightened medical and pharmacological sophistication, it is now increasingly acknowledged that the widespread faith in plant medicinals is not founded only in superstition. Many plants do contain substances that produce physiological effects, and which may be valuable in combatting disease agents. In a given botanical remedy, there are usually numerous pharmacologically active ingredients. Some have similar qualities and are synergistic, whereas others are antagonistic. Of course, it might also be argued that merely believing in the value of an herbal remedy may have some therapeutic significance.

We do know that specific plants have been associated with certain ailments for millennia. At least several hundred drug plants were known to the Assyrians more than 4000 years ago. The earliest plant extract used medicinally may have been gum arabic, derived from various species of *Acacia* (Leguminosae). It was mentioned by Herodotus as being of medical interest before the fifth century BC, and today it is still in use, although primarily as an emulsifier in cough syrups and tinctures. And morphine, derived from opium, has long been used as a pain killer.

In Greece, Dioscorides (about 75 BC) considered the medicinal value of several thousand plants in his *De Materia Medica*, a work that became a standard pharmacological reference for the following 1500 years. Then, in the 15th through the 17th centuries, the proliferation of herbalists preceded the growth of modern botany and pharmacology. During the Middle Ages, the Doctrine of Signatures asserted that all plants had been created for human use, and that their shapes indicated effectiveness in treating particular areas of the body. A walnut, for example, was believed to be beneficial to the brain. (Note the similarity of this concept to the beliefs of the Chippewas.) Even today, wild plants are sold as medicinals through respectable outlets. More than 125 species that are collected from the wild in Appalachia are for sale through drug houses, and a recent list of medicinal plants still being used by the Maori of New Zealand numbers well over 100. From the most primitive to the most sophisticated cultures, botanical medicines continue to be prevalent.

Origins of Western Pharmacology   Western pharmacology can be most clearly traced in origins to the river valleys of the Nile, the Tigris, and the Euphrates. The influence of practices in these areas spread with transcultural contact into Europe. In 1240, the German Emperor Frederick II issued an edict that may be considered the Magna Charta of Western pharmacy. Although it originally applied only to a limited area known as the kingdom of the two Sicilies, the tenets of the edict were adopted throughout Europe during the next several centuries. There were three essential regulations:

1. separation of pharmacy from the medical profession;
2. official supervision of pharmaceutical practice;
3. obligation by oath to prepare drugs reliably, according to the skills of the art, and in a uniform and suitable quality.

These basic ideas remain integral to modern pharmacology, which in the United States is under the control of the Pure Food and Drug Administration.

As medicine became more sophisticated during the 1800's, experimental pharmacology made progress as well, and the foundations of contemporary Western pharmacy were firmly established. The practice of pharmacy began to achieve distinction in the United States when the first *United States Pharmacopeia,* compiled by Dr. Lyman Spaulding, was published in 1820.

Oriental Medicine   Oriental medicine, which has followed some courses distinctly different from the West, also has employed numerous plant products. Licorice root, for example, has a long history of use, and has recently been substantiated as an effective medicinal. It tends to depress and ease pain and also increases the effects of other drugs. Apparently it is a cure for ulcers (Tagaki, et al, in Chen, 1965). Ginseng extracts, considered in some Oriental cultures to have worth as everything from aphrodisiacs to sustainers of youth, have been shown to have a stimulating effect on the cerebral cortex (Petkov and Staneva-Stoicheva, in Chen, 1965).

Many pharmacological products are effective due to the presence of alkaloids, which are impressive in the physiological responses they cause in animals. Although, today, a variety of pharmaceuticals are chemically synthesized, these drugs are patterned after natural products. Pharmacy currently constitutes a major business in the United States, where more than 3 billion dollars were spent on such products during the late 1960's.

Suggested Readings

Aaronson, Bernard, and Osmond, Humphry. *Psychedelics.* New York: Doubleday and Company, Inc., 1970.

Adams, Francis. *The Seven Books of Paulus Aegineta.* 3 Vols. London: The Sydenham Society (Translated from Greek), 1844.

Adams, Richard Newbold. *Energy and Structure.* Austin: The University of Texas Press, 1975.

Allegro, John M. *The Sacred Mushroom and the Cross.* New York: Doubleday and Company, Inc., 1970.

Castaneda, Carlos. *The Teachings of Don Juan: A Yaqui Way of Knowledge.* Berkeley and Los Angeles: University of California Press, 1968.

——. *A Separate Reality.* New York: Simon and Schuster, 1971.

——. *Journey to Ixtlan.* New York: Simon and Schuster, 1972.

——. *Tales of Power.* New York: Simon and Schuster, 1974.

Chen, K. K. and Mukerji, B., eds. *Pharmacology of Oriental Plants.* New York: The Macmillan Company, A Pergamon Press Book, 1965.

Colle, Henry. *Heritage of Pharmacy in Asia.* Copyrighted by Henry Cole, 1945. Unpublished Manuscript.

Densmore, Frances (1974). *How Indians Use Wild Plants For Food, Medicine, and Crafts.* New York: Dover Publications, Inc. (This book is a reprint of an article in the 44th Annual Report of the Bureau of American Ethnology; 1926–1927).

Harner, Michael J., *Hallucinogens and Shamanism.* New York: Oxford University Press, 1973.

Hesse, Erich. *Narcotics and Drug Addiction.* New York: Philosophical Library, 1946.

Janick, Jules; Schery, Robert W.; Woods, Frank W.; and Rutton, Vernon W. *Plant Science*. San Francisco: W. H. Freeman and Company, 1969.

Kramer's and Undang's *History of Pharmacy*, revised by Glenn Sonnedecker, Philadelphia: J. B. Lippincott Company, 1963.

Lewis, Oscar (1955). "Medicine and Politics in a Mexican Village," in *Anthropological Essays*. New York: Random House, Inc., 1970.

Schenck, Gustav. *The Book of Poisons*. Translated from German by Michael Bullock. New York: Rinehart, 1955.

CHAPTER

16

ETHNIC NUTRITION

Almost any cultural group has an ethnic diet peculiar to its heritage. Why are certain traditional foods and combinations of foods eaten? Corn and beans are the typical diet of much of Latin America (see Chapters 6 and 7). People in Latin America have eaten corn tortillas and beans since 3000 BC. Archeologists found proof of this in samples of crop plants and grinding implements uncovered at Tehuacan in Mexico. This diet is also usually supplemented by chilies, hot peppers of the *Solanaceae* found throughout Latin America. In many eastern Mediterranean cultures, legumes such as beans and chick peas are traditionally eaten with wheat products. In Scandinavia, peas are served with rye or wheat. The Orient has a heritage of rice and soy bean products such as soy sauce, soy bean curds (*tofu*), and other soy bean foods. Few cultures are as limited in their diets as the ethnic fare would suggest, yet there is usually a sound basis for the origin of culturally defined diets.

The contemporary diet in the United States, although it may be criticized as nutritionally inadequate relative to the cost for consumers, is notable in two respects. In composition, it is reasonably cosmopolitan; and it is transseasonal—the majority of fruits, vegetables, and meats being available year-round. Also, our diet is characterized by a greater quantity of animal protein than diets in other parts of the world.

The Changing Human Diet   As evidenced by changes in dentition and related changes in jaw and skull structure, early human beings had a marked tendency toward an omnivorous diet. Presumably, this was true from about 500,000 years ago until the end of the Pleistocene Epoch. Subsequently, the agricultural transition began and plants became an increasingly important element in the human diet.

We know that during recent historical times there was more of a demand for animal protein than could be met by the supply. Only the wealthy and those of high social classes ate much meat. The great majority of people generally satisfied their caloric and nutritional needs predominantly with plants. During the Middle Ages in Europe, animal protein was so rare a commodity that it was eaten only on feast days and other special social occasions. Of course, there are always exceptions to so broad a generalization: the peoples of Arctic lands and nomadic groups stand out in this respect. In some cultures, domestic animals were regarded mainly as beasts of burden. Possibly the animals were not reproductively prolific enough to be a regular source of food. In areas where the agricultural lifestyle predominates, plant materials formed the basis of most diets, for most of the people, most of the time.

Today, because of cost, land use priorities, and other factors, it is evident that animal sources cannot be expected to supply adequate protein for most of the world's peoples. Until recently, agricultural optimists predicted abundance for the whole world. Now, it is clear that even the affluent may have to get along with less. An understanding of the nutritional basis of ethnic diets, which are primarily vegetable proteins, is essential if we are to deal with the world hunger problem. We must comprehend why such diets are sufficient if we are to improve plant diets for the future—both in the developed and in the developing countries.

If we imagine all organisms in food chains, the efficiency of energy conversion is roughly 10% at every point of transfer. Thus, when we eat the protein of animals that have consumed plant materials, the diet is less efficient than if we used plant proteins directly.

Human Protein Needs Protein is necessary for the maintenance and growth of all living organisms; human beings are no exception. A human is about 18% to 20% protein by weight. Protein constitutes not only structural tissues found in organs and muscles; it makes the enzymes for metabolism, the hormones required for physiological regulation, and the body fluids such as blood. Any diet to be adequate must provide a sufficient quantity of protein, and the appropriate combination of amino acids as well.

Amino Acids All proteins are not created equal, but differ significantly according to the source. Proteins comprise about 20 amino acids; some of these cannot be synthesized by our body tissues and must be obtained from external sources. For human beings, here are the *essential amino acids*: tryptophan, leucine, isoleucine, lysine, valin, threonine, methionine, and phenylalanine. In addition to these, histidine and arginine may not be as readily synthesized by children as by adults. For growth and maintenance, then, eight amino acids are needed in the diet continuously and simultaneously. If one is missing or present only in low levels, protein synthesis is reduced, or may stop entirely. In addition, the essential amino acids must be present in a specific ratio. Most foods contain all eight, but in the vegetable proteins, which comprise so many ethnic diets, one or more will usually be found only in small quantities. Let us assume that the amount of one amino acid is inadequate in such a diet. It would then be the *limiting amino acid* for the metabolism of the food material.

An inadequate quantity of one amino acid is considered limiting because the remainder of the amino acids of the essential series will be metabolized *only at the level allowed by the one present in least quantity.* In the synthesis of protein by our body cells, the rate and level of synthesis is dictated by the least abundant amino acid.

Another factor in amino acid metabolism may be called *net protein utilization.* This term refers to the amount of protein actually absorbed which is made available to the body for protein synthesis. The amount of a food available for net protein utilization depends on how closely the ratio of the eight essential amino acids in the food corresponds with the ratio of the proteins in human cells. The protein of egg is often taken as a model for the amino acid ratio of human beings.

Quantitative Requirements   Beyond these qualitative protein requirements, there is a gross quantitative requirement as well. In order to maintain body tissue, the minimum protein requirement is from 0.214 to 0.227 grams per pound of body weight per day. These figures become meaningful when we consider how much of certain foods must be eaten in an ethnic diet to provide this basic nutritional requirement. In evaluating the nutritional potential and efficiency of an ethnic diet, nutritionists must establish protein allowances based upon the net protein utilization of the common proteins in that diet:

$$\frac{\text{Protein allowance (based on egg as a standard)}}{\text{Net protein utilization characteristic of the ethnic diet}} = \frac{\text{of 0.27 grams/pound of body weight} \times 100}{}$$

total grams of protein for ethnic population
per person per pound of body weight per day.

Animal proteins provide a greater number of the essential amino acids than do plant proteins. In a diet in which the protein is obtained solely from animal flesh, the human requirement is about 0.40 grams of protein per pound of body weight per day. In contrast, if plants are the protein source, the figure is about 0.51 grams per pound per day. Successful ethnic diets have complementary foods which combine plant proteins in such a manner as to reflect the ratio of human beings' amino acids. Such a result is usually not achieved consciously, but on analysis, it is the reason for the sufficiency of many ethnic diets. People undoubtedly created such diets by experimentation, even though they did not call it that. The efficacy of ethnic diets constructed around the use of vegetable matter is reflected in the health of the population as a whole; for protein deprivation during early childhood can limit the mental development of an individual for the rest of his or her life.

Cultural Staples   Any food that is more than 10% protein by weight (regardless of the quality and net utilization of the protein) can be classed as a *cultural staple*. Foods in this category are wheat, soy beans, many nuts, legumes such as peas and beans, certain strains of corn and rice, and some others. To compensate for the lesser quality of plant proteins, adequate combinations of plant foods are essential. These *compensatory plant proteins* are notable in one respect: the combination is often greater than the sum of its parts.

Legumes and Grains   Legumes are cosmopolitan, indigenous to both the New and Old World, and have been culturally important for thousands of years. Considered in terms of the essential amino acids, legumes are strong in isoleucine and lysine, but weak in tryptophan and sulphur-containing amino acids. By contrast, grains, such as corn and wheat, are deficient in isoleucine and lysine, but rather high in other low-level amino acids absent from legumes. In combination, legumes and grains supply complementary proteins—increasing the net protein utilization. Thus legumes and rice, beans and wheat, and beans and corn are nutritionally efficient cultural solutions to obtaining enough protein.

A Successful Mesoamerican Diet   Dietary combinations must do more than meet protein needs. The lack of vitamins or minerals may result in a variety of diseases. *Scurvy*, a disease characterized by anemia, weakness, spongy gums, and bleeding from mucuous membranes is caused by a deficiency of vitamin C. *Pellagra*, characterized by skin eruptions, nervous disorders, mental deterioration, and other symptoms, results from niacin deficiency. Pellagra, while rare today, once plagued peoples in various parts of the world who were bound to diets, by custom or necessity, lacking in niacin. In areas where corn was a staple, pellagra was often common. But among Mesoamericans, it seems to have been rare due to the development of a complementary diet. Corn was the staff of life and civilization in the New World, but the traditional accompaniment of beans and, especially, chilies, provided niacin. Also, the manner of preparation of corn tortillas is significant. Prior to being ground, the corn grains are mixed with slaked lime and water; then heated for 18 hours or more. This process aides in the removal of the hulls. The Indians who originated this procedure were not aware of any dietary

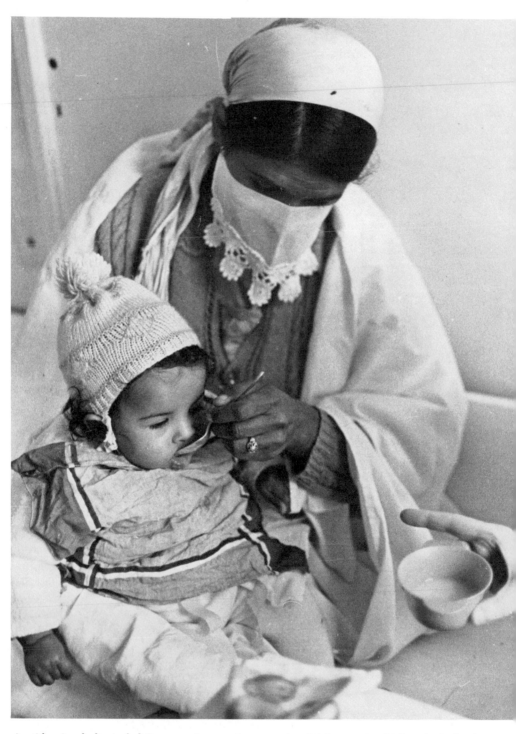

An Algerian baby is fed Superamine, an inexpensive, high-protein, high-calorie food mixture based on indigenous foods. (FAO photo)

benefits, but the alkali-treated corn contained far more niacin than untreated corn, because the lime releases chemically-bound niacin from the kernels. These aspects of the Meso-american diet were not achieved intentionally. The Indians did not know about niacin, nor did they have any reason to be consciously aware that particular food combinations were nutritionally expedient. "Synergistic combinations" (Kaplan, 1967) of foods have evolved, in part, because they do satisfy a population's needs, but the development of such combinations is a multivariable "accident." Given different circumstances, equally adequate diets of different kinds might have evolved.

Middle Eastern Diets   In the Middle East, diets are based mainly on grains (Sabry in: *National Academy of Sciences*, 1961), which provide more than 70% of the average daily calorie supply. In Middle Eastern diets meat, eggs, fish, and milk products constitute slightly more than 6% of that total. Most of the available protein comes from cereals, pulses, and particularly, wheat in the form of bread. Maize is also made into a bread, especially in Egypt, and rice is important. Lentils, chickpeas, broad beans, and sesame are used in the preparation of many ethnic dishes. Of the Middle Eastern countries, Lebanon has a relatively high standard of living, evident in a higher per capita consumption of animal proteins; but even there kwashiorkor (severe malnutrition) is apparently a frequent condition.

Protein malnutrition is not uncommon among the lower classes in all Middle Eastern countries. The diet of grains and pulses may supply a sufficient quantity of protein, but quality is a problem, especially among the poor, whose diet may be limited largely to bread. Such a diet does not have an adequate complement of cereal and pulses. Middle Eastern bread is flat and round; like the tortilla in Latin America, it is eaten throughout the Middle East. Combined regularly with pulses and occasional animal products, the diet should be fairly adequate. But the poor cannot always meet even this minimal need. A few additional proteins are obtained from the leaves and stems of wild plants which are eaten in quantity by people in Middle Eastern countries. The plants may be consumed raw, or cooked with rice or parboiled wheat. The protein content is minimal, and little is known of the total nutritional value.

Diet in Africa   In Africa, the manifestations of kwashiorkor are commonly present among the poor. Malnutrition is a result not only of dietary deficiency, but other associated causes: respiratory infections, infestations of parasites, diarrhea, and others (Jelliffe, 1961). Often, protein-calorie malnutrition occurs when there is a migration from tribal units into the slums of cities or other settings in which social disintegration and related problems occur. It was found that the Watindiga, who are hunters, and the Acholi, who subsist largely on millet, were basically free of malnutrition problems in their native settings. Families of the Acholi who had moved to the city of Kampala to work had changed their diet of millet to one of corn and plantains. The children of these people suffered from malnutrition.

On the other hand, the Baganda tribe, a more culturally sophisticated group, has a higher rate of malnutrition, due, primarily, to their acculturation to Western society. The period of breast feeding has been shortened; infants are bottle-fed earlier. This has resulted in nutritional loss and the risk of infectious diarrhea. There is too much emphasis in the diet on plantains, which are low in protein. Lastly, heavy infestations of hookworm contribute to the state of malnutrition, which in any country is often the summation of various factors besides diet. Parasites are a prevalent problem in tropical areas. These organisms consume the energy and nutrients of the host, and, under the worst conditions, cause extensive physical damage.

Corn is not culturally intrinsic in Africa as it is in Latin America. However, it is now a staple food in many areas of the continent. In Usambara, for example, corn is present in about 27% of all breakfasts and 80% of all lunches and dinners (Schlage in: Kraut and Cremer 1969). Corn is eaten there as a porridge or made into a thin soup, and three kinds of flour are made from it: *dona*, a commercial whole meal preparation; *wnga*, a flour prepared in the home; and *sembe*, a flour available from different milling industries.

Rice is also eaten, but not as much. *Mandazi*, a doughnut made of wheat flour and yeast, is deep-fried in fat and commonly eaten as a snack. Cassava, the second most important staple, is eaten in times of maize shortages. The tuberous roots are peeled and boiled or roasted on an open fire. A flour made from the cassava root is usually eaten as a porridge. Bananas, which have a protein content of less than 2%, are popular. There are at least 10 varieties in Usumbara; one is used for cooking, and the others are eaten fresh. In the highlands near Kilimanjaro, bananas are the main staple, an unfortunate choice in view of their negligible protein content. Sweet potatoes are sometimes eaten, but do not have a major dietary role. Beans (*Phaseolus vulgaris*) are the most important source of protein next to corn, providing 10% to 30% of the total diet. Smoked fish is another popular food, and is available in markets throughout the year. The fish is an important source of protein; but due to the relatively high cost, it is eaten in small quantities by most of the population. When dry, the smoked fish is somewhat more than 40% protein.

As can be seen from this listing of typical foods for a specific area, the potential for an adequate diet is often present; yet cultural and economic factors limit the potential. If everyone in Usumbara were to have a daily intake of corn, fish, bananas, and beans, accompanied by various garden vegetables, nutritional requirements would be satisfied. But the realities of life make such a diet difficult to achieve. The smoked fish, which is the best source of protein, is expensive. Also, if it were made available in quantity, the supply would be soon exhausted. The emphasis on corn is offset by the addition of beans, which combine to form an acceptable protein, as in Latin American diets. Even so, the majority of culturally intrinsic foods are starchy and low in protein. Without the fish and legumes, the diet would probably be unacceptable.

In a typical market in Usumbara, these foods are available throughout the year: bananas, cabbages, cassava, coconut, maize, papayas, plantains, sweet potatoes, wild spinach, sugar cane, taro, tomatoes, smoked fish, and salted fish. Available year round but in less regular quantities were French beans, kidney beans, lemons, onions, oysternuts, green peppers, pineapples, Irish potatoes, and some other fruits and vegetables. There are also vegetables and fruits that appear seasonally, such as cowpeas, mangoes, and passion fruits, and other foods that are only occasionally offered, such as custard apples, pears, and white ants (termites). It would seem, from this selection, that a nutritional balance would not be a problem. But again, economics and cultural habits affect the composition of ethnic diets.

The Effects of Malnutrition   Perhaps the most severe effects of malnutrition are those occurring in infancy and early childhood. Experiencing malnutrition in the early years may scar an individual for life, by causing inferior development both of the brain and of the physique. In a group of children studied in a village in the Guatemalan highlands, a high infant mortality rate, (83 per 1000 births), a high mortality rate of children in the age group of one to four years (54 per 1000), and also a high birth rate (50 per 1000 of the general population) were observed (Mata et al. in: Wolstenholme and O'Connor, 1967). Food shortages are a common occurrence, and cultural practices are also responsible for a low intake of foods containing high-quality proteins. Also, the environment is not conducive to good hygiene, a factor adding to the consequences of inferior nutrition.

In comparing these children with children in Iowa of the same age,* it was found that at the age of six months, the Guatemalan children began to exhibit physical and mental deficiencies, and at the age of nine months, 68% of them manifested deficiencies.

Relevant to this is that the Guatemalan infants were breast-fed up to the age of six months. Occasionally they were given small quantities of sugar water, rice water, and diluted coffee. By one year of age, the children were being given small amounts of a dietary supplement much like the adult fare: black beans, tortillas, some meat, and vegetables. The nutritive value of these foods was equal to approximately 17% of the protein and 15% of the calories needed daily by the children for optimal growth. From the age of six months, the value of mother's milk becomes increasingly less sufficient. In this particular diet the supplements are nutritionally inadequate and also are a source of infections. Entero- and adeno-viruses were common from the first week of life on, and with the feeding of more supplemental foods after the age of six months. The presence of *Shigella* and *Salmonella* bacteria was evident; the former is invariably associated with diarrhea. Correlations were found in the two populations between growth rates, malnutrition, and infection—suggesting a relationship between poor nutrition and susceptibility to infections.

Not only is an infant's physical and mental maturation affected by the quality of diet during the early years, but the nutrition of the mother influences the developing fetus. If a mother's diet fails to provide her own protein and nutrient needs, fetal growth will obviously be affected. So, also, will the quality of the maternal milk with which she nourishes the baby when he or she is born.

A nutritional study of Kikuyu farmers in Kenya revealed that their combination diet is also based on the complement of maize and beans. This supplied them with most of their calories, protein, iron, and B vitamins. The average nutrient intake, considered in terms of the adequacy of quantity rather than quality, against a standard of 100%, provided: 94% of needed calories, 164% of protein, 44% of calcium, 169% of iron, 28% of vitamin A, 485% of thiamin, 143% of riboflavin, 126% of niacin, and 89% of vitamin C (Keller et al. in: Kraut

This standard was used because previous research has shown that Guatemalan children who are well nourished mature at about the same level as children in Iowa.

and Crener 1969). The diet was found to be sufficient but not optimal, particularly in falling short of the calories needed. Dietary adequacy was shown to be directly related to total cash income; it is probable that those earning higher incomes were able to provide themselves with more meat.

The composition of ethnic diets depends, to some degree, on the necessity of fulfilling basic nutritional demands. A reasonable assumption is that, if a customary diet is too severely wanting in essential nutrients, the people eating it have to make dietary changes or die off. However, the fact that a given diet is workable does not mean that it is optimal. As we have seen, cultural groups may lead somewhat marginal existences on foods that supply enough nutrition to live but provide less nourishment than is needed for good health and a satisfactory life quality. Some ethnic diets may be culturally ingrained for centuries. Whether in Guatemala, Africa, India, or elsewhere, altering these traditional patterns is more easily proposed than done. In India for example, the Hindu religious proscription against eating beef (chicken, mutton, and fish are less frowned upon) for a majority of the population aggravates a situation of extensive malnutrition.

The Upgrading of Local Diets   A significant worldwide effort is being focused on the problem of upgrading local diets as well as just providing enough food. Researchers have devised various combinations of plant proteins which supply the essential and additional amino acids. Usually these protein supplements are in the form of flour that can be added to or substituted for the customary ingredient, or in the form of powders that can be made into milk-like drinks.

A New Kind of Flour.   In Latin America, the Institute of Nutrition of Central America and Panama (INCAP) developed a vegetable flour, *Incaparina* (from INCAP and *harina*, the Spanish word for flour), which is produced largely from cottonseed meal. It has 25% or more protein, and costs about a fourth as much as milk. Apparently, much of the sale has been to middle-class families instead of to the poor who need it— probably because the latter are more reluctant to accept innovative changes in diet. INCAP has experimented with a variety of vegetable mixtures in the attempt to produce a low-cost, high-protein, easily-synthesized meal. One such meal contains 28% corn, 28% sorghum, 38% cottonseed flour, 3% Torula yeast, and 3% leaf meal.

Peanut Flour.   Another potential flour substitute of high nutritional value is peanut flour. Preparations of peanut flour

Above *A mobile demonstration unit helps Kenya women visualize physical benefits of increased milk consumption. (FAO photo)*

Left *A nutrition expert explains the proper method of preparing food from protein-rich rations in an Andean village in Chimborazo province.*

fed to infants in a mixture of water and sugar were accepted satisfactorily, and digestion was good as long as the amounts did not exceed an average of 50 to 80 grams per day (Senecal in: *National Academy of Sciences*, 1961). Experiments have indicated that the peanut flour alone may not be adequate, especially for persons already suffering from protein deficiency. However, if the flour is combined with an animal product, such as fish flour, the nutritional value is increased. A millet-peanut mixture is also quite effective.

Nitrogen: A Nutritional Necessity.   The quantity of nitrogen present in a food material is related to the material's nutritional adequacy (Brock in: *National Academy of Sciences*, 1961). The indication is that a diet of lesser protein value can be sufficient if the level of nitrogen intake is high. When nitrogen is limited, the differences in nutritional worth of different foods become far more evident. If the quantity of nitrogen present is very low, it will not be possible to ingest enough food to supply minimal requirements. The "pot belly" characteristic of children who are the victims of malnutrition is explicable in terms of nitrogen deficiency. In unconsciously trying to satisfy the body's demand for more nitrogen, a child eats enough of the low-protein staple, whether cassava, corn, or another food, to distend his or her intestines. The condition is then further aggravated by the constipating effect of some staple foods.

Introducing Alternative Diets   In spite of the cultural barriers erected against changes in traditional diets, it is not an impossible end to accomplish. In terms of the local economics of an area, it is necessary, before introducing new foods, to calculate the cost per unit quantity of nutrients. That is, if a given food which is higher than another in protein content costs four times as much to produce, it may not be a feasible dietary alternative.

Let's look at an example. Analysis proved beans to be the least expensive protein source in Usumbara. Maize was the next cheapest. These are already staples, so the nutritionists might direct efforts toward an increased use of beans instead of the introduction of new crops. After beans and maize, the next most inexpensive sources of protein were found to be oyster nuts and spinach. Although oyster nuts are, at present, eaten almost exclusively by lactating mothers, since the nuts are high in protein and are relatively inexpensive, this food might advantageously be produced in greater quantities. It is interesting that a high-protein food such as oyster nuts and a low-protein vegetable such as spinach cost about the same per unit of protein. One would have to eat prodigious quantities of spinach to ingest much protein; but as the vegetable has a high mineral value, it is a good accompaniment for high protein foods.

Soybean Products   In the future, products made from soybeans may be a significant factor in combatting protein deficiency (see Chapter 9). Many Japanese and other Orientals thrive on diets containing little or no animal materials. They derive their protein predominantly from soybean products, such as *tofu, miso,* and *natto. Tofu* is a curd, and the latter two are bean pastes made by fermentation with microorganisms (*Aspergillus oryzae* and *Bacillus natto*). In *natto,* all of the essential amino acids, except threonine, are increased (Sano in: *National Academy of Sciences,* 1961). The protein content of *miso* is lower than in other soybean products. Even so, it resembles skimmed milk powder in its complement of amino acids. *Natto* and *miso* keep well, an advantage in distributing protein foods to the undernourished in developing countries.

Due to fermentation, the content of essential amino acids in many soy products, such as *natto,* is higher than the protein content of soybeans. Experiments with rats indicate that skim milk (and any animal product) is probably a better source of protein than soybeans and soybean products, but *natto* and *miso* are nutritionally almost on a par with the milk. Lysine, usually added to cereals to increase their quality, is present in a nearly sufficient amount of soybean foods, which would be better fortified with methionine and cystine. During fermentation with *Bacillus natto,* water-soluble nitrogen increases. This enhances the value of the protein. The content of vitamin B_2 also rises (Arimoto, in: *National Academy of Sciences,* 1961). An Indonesian soybean product, *tempeh,* may also provide a quality protein source.

Other foods which may help solve the worldwide problem of protein deficiency include: Mysore flour, a blend of peanut and tapioca flours; *paushtik atta,* made from wheat flour, peanut flour, and tapioca flour; malt foods and substitutes for dried milk synthesized from peanuts and soybeans. Further research is certain to result in a more extensive and more palatable array of quality protein foods created from plants. Consumers are often motivated by habit. Accordingly, attempts are underway (see Chapter 9) to manufacture foods, made from vegetable proteins, which resemble meat products. These considerations are relevant to affluent countries, and are secondary to the production of digestible, quality protein which will be acceptable and affordable to the inhabitants of developing nations. Although scarcity plays a key role in the crisis of global malnutrition, other factors may have as great an impact.

Learning from Ethnically Evolved Diets   There is some disagreement over the question of whether or not complementary combinations of plant proteins are nutritionally comparable to animal products. In the context of the world hunger problem, the issue is essentially unimportant, simply because it will not be possible for a significant percentage of the human population to be sustained on diets containing quantities of meat. So, if plant foods are second best, plant materials must nonetheless be accepted as the functional answer for feeding the expanding numbers of human beings. Ethnically evolved diets, which may be flawed in terms of ideal nutrition, offer insights into the means for solving the world food crisis with minimal energy expenditure. By working within the framework of culturally accepted diets, there will be less difficulty in introducing new foods or in emphasizing the importance of combinations different from those currently preferred. The quest to realize quality as well as quantity in meeting human food needs requires tactical as well as technological awareness.

Suggested Readings

Kaplan, L. "Archeological *Phaseolus* from Tehuacan," in *The Prehistory of the Tehuacan Valley; I. Environment and Subsistence.* Edited by D. Byers. Austin, Texas: The University of Texas Press, 1967.

Kraut, H., and Gremer, H. D., eds. *Investigations into Health and Nutrition in East Africa.* Munchen, Germany: Weltforum-Verlag, 1969.

Mata, Leonardo J.; Urrutia, Juan J.; and Garcia, Bertha. "Effect of Infection and Diet on Child Growth Experience in a Guatemalan Village," in *Nutrition and Infection.* Edited by G. E. W. Wolstenholme and M. O'Connor. Boston: Little, Brown, and Company, 1967.

Mayer, Jean, ed. *U. S. Nutrition Policies in the Seventies.* San Francisco: W. H. Freeman and Company, 1973.

National Academy of Sciences National Research Council. *Meeting Protein Needs of Infants and Children.* Washington, D.C.: National Academy of Sciences, 1961.

Roe, Daphne A. *A Plague of Corn: The Social History of Pellagra.* Ithaca, New York: Cornell University Press, 1973.

Wolstenholme, G. E. W., and O'Connor, Maeve, eds. *Nutrition and Infection.* Boston: Little, Brown, and Company, 1967.

During the past decade, food shortages have been dramatically accentuated. A new awareness began to develop, at about the same time that Paul Ehrlich's *The Population Bomb* drew the public's attention to the real threat of overpopulation. Not only is our Cornucopia finite, but pressing food scarcities may be imminent and not just a phantom of the future. Now, most of us are quite aware that this is true. Our final unit begins with a consideration of the "Green Revolution," which provided so much hope for future agricultural abundance. We then examine some of the faults inherent in modern agricultural practices and in plant breeding methods. We find wanting the idea that any substantial solution to food shortages may come from the oceans. Chapter 21, on innovative solutions, is intended to offer possibilities that may partially alleviate shortages, not to promote radical changes in our eating habits. The final chapter is an overview of the current world food situation.

UNIT 5

CHAPTER

17

THE GREEN REVOLUTION

The initial phase of the current Green Revolution began with a 25-year breeding program instituted in Mexico under the auspices of the Rockefeller Foundation. Wheat strains were bred for resistance to rust, and to produce higher yields. Cross-breeding with short-stemmed Japanese varieties (Norin) produced a high-yield strain (Gaines) which grows very well in the northwest United States. Varieties with high yield and rust resistance were developed in Mexico, but required more irrigation than the ordinary varieties.

In the Philippines, the International Rice Research Institute was established in 1955. Devoted to the breeding of rice strains that would give higher yields and resist fungal and insect attack, the Institute developed a strain (IR-8) that is especially productive. It originated from a Taiwan breed that had precursors in Indonesia, Ceylon, the Philippines, and other eastern locales. In both wheat and rice, short stems are considered necessary for high productivity. Varieties with long stems tend to bend under the weight of a high yield. Also, breeding was directed toward reducing photoperiodic sensitivity. By breeding to diminish the influence of day-length, the plants could then be productive in a variety of geographical latitudes with less daylight.

Increases in crop yields as a result of using improved strains were impressive and rapid. India's wheat harvest was 35% higher in 1968 than in the previous year. The increase was largely because of the introduction of improved varieties. In Pakistan, the production was 37% higher. For the same reasons, rice production in Ceylon and in the Philippines soared between 1966 and 1968. In India, in Mysore State, which is one of the less developed areas of that country, three corn crops are now possible every 14 months. Two or three rice crops are grown annually in India, Indonesia, and the Philippines because of improved irrigation and better varieties of rice. Favorable weather conditions were also a factor in this increased output, but the introduced strains were a significant cause.

Can the Revolution Meet Our Needs? If it were possible to initiate the production of these high-yielding varieties of grain into a world with a relatively stable human population, then the most optimistic projections of proponents of the Green Revolution might be realized. However, this cannot be done. The human population is not stable, and "miracle crops" cannot free human beings from the grip of hunger and malnutrition. Certainly there are factors other than population growth—factors such as economics, transportation and storage—which are influencing the situation. However, we feel that overpopulation is the basic underlying cause of the problem. When human numbers are projected to double within the next thirty years, it is unrealistic to minimize the overwhelming role of overpopulation. If it is not possible to provide for those already living, where will we get the food for more *if* the population growth rate is not significantly reduced?

Implementing the Revolution   The Green Revolution has a potential for a partial means of increasing food supplies. Innovative plant technology alone has not carried the revolution. The use of fertilizers, pesticides, and increased irrigation usually have to be accelerated for the new strains to be effective. A debt is owed to the scientists who developed better fertilizers, pesticides, and the techniques for their application, as well as improved methods of irrigation. (Some serious disadvantages resulting from these methods will be analyzed later in the chapter). For example, approximately half of the world's rice crop was destroyed annually by fungi and insects before better pesticides were implemented. New rice strains may be resistant to one kind of parasite, but such strains may prove to be even more vulnerable to attack by other pests. A fungus and a stem borer were found to be especially attracted to new rice varieties cultivated in India and Ceylon.

Even were there no intrinsic problems associated with the growing of engineered crop plants, the introduction of these strains is not as easy as one might suppose. The new crop may not be as desirable to the local populace, and may command a lower price in the markets. Such a situation will encourage farmers to continue the cultivation of traditional varieties, even if the yield is lower. Another discouraging aspect of the Green Revolution is that the developing countries, which stand to benefit the most from technologically improved grains, often do not have researchers and technicians. Organizations such as the International Rice Research Institute in the Philippines and the International Maize and Wheat Improvement Center in Mexico should be fostered in as many developing nations as possible, even if the already developed countries must bear much of the supporting expense. Hopefully, this would be a more useful course of action than the practice of shipping grain to countries that cannot feed their populations—assuming that the latter course could be continued indefinitely, which it cannot. With such a research organization actively involved in each country, research goals could be instituted to meet a given nation's particular set of agricultural difficulties and requirements such as climate, terrain, pests, and other factors.

Above left *From existing breeds of rice, India's Central Rice Research Institute creates higher-yielding ones which are better suited to local conditions, and are more resistant to disease. (FAO photo)*

Above right *An agricultural plant geneticist innoculates the stem of a wheat plant to induce an epidemic of wheat rust. Scientists then select plants that resist the rust. (FAO photo)*

Left *An agricultural geneticist assesses the productivity of a new rice hybrid. (FAO photo)*

Analyzing the Results   Unfortunately, statistics supporting the impressive success of the Green Revolution can be misleading without complete information on all related circumstances. As an example, consider the yield increase cited for India from 1967 to 1968. The latter was a wetter year, and even a harvest of the traditional wheat might have been expected to be greater. Presumably, much of the gain was due to the use of higher-yield grains, but there is of course no way to measure how much of that increase is due to weather and how much to the new strains. Misconceptions may also result from figures giving yield increases unless such figures are compared with the amount of land under cultivation from one year to the next. As the population grows, more land will usually be cleared for cultivation. Unless the acreages being farmed in successive years are known, it is not possible to determine the percentage of yield that is a result of new strains and the percentage that is a result of more land being farmed. The patterns of cultivation are an important part of this whole picture. If irrigation techniques are refined, it may be possible to produce, annually, two or even three crops where previously only one could grow.

If such facts are not known, a marked increase might be attributed to high-yield crops when, in fact, the increase is a product of irrigation. Finally, a new strain's potential must be tested longer than a couple of years. If the yield remains relatively high over a period of climatic fluctuations and other varying conditions, then the assumption of a strain's superiority is probably valid. As William Paddock underscored, (Paddock, 1970), weather and other extenuating conditions are too infrequently considered in the official assessments of crop successes. If, in a given year, the yield is low, the weather is usually blamed; if the yield increases, success is often attributed to the effects of technological innovations and other factors for which governments and developmental agencies claim credit.

Even in the most favorable light, the Green Revolution has had considerably less impact than the word *revolution* implies. Let us consider India as an example. About a third of the population of Asia lives there. Of all the new strains of various crops, only the new strains of wheat have gained importance, and wheat is a less significant crop than rice in that country. The new hybrid rice varieties have had minimal success there.

Another fact to consider in analyzing the Green Revolution's successes is the relatively small percentage of the world that has been affected by the revolution. Lowell Hardin of the Ford Foundation referred to the area as "a postage stamp on the face of the earth" (Hardin, 1969). Why have the new grains not spread throughout the world's hunger zones? A primary reason is the logistic and implementation problems discussed previously. Irrigation is a necessity for the successful cultivation of the high-yield crops. This automatically excludes many regions. Much of the earth's land surface is dry, and irrigation is expensive. Also, the technology for irrigating vast areas of farmland simply does not exist in many countries. Even if it did, the wisdom of turning deserts into gardens is disputable (see Chapter 22). Again, we are forced to confront the role of overpopulation. Dry lands can sustain limited human populations if there is a balance with the intrinsic capacities of the environment. It is when human numbers overflow these limiting parameters that longterm food crises arise.

In tropical countries relatively little research has been carried out to investigate increasing the productivity of tropical crops. The Green Revolution is based on wheat and rice. The former, in particular, is not as suited to tropical agriculture as other plants might be (Tydings 1970). Largescale agricultural research stations located in every tropical country should be established, even if much of the support for such stations must come from developed countries in the Northern Hemisphere. Rather than concentrating on the perfection of a few crops, the emphasis might be better placed on the improvement of numerous high-nutrition foods which grow well and which are culturally accepted in a given country. Organizations such as the Philippines International Rice Research Institute and the International Maize and Wheat Improvement Center in Mexico are of great value in the effort against hunger. However, their research should be supplemented with other investigations into enhancing the potential of indigenous crops.

Socioeconomic factors certainly influence the Green Revolution. The Philippine Rice and Corn Administration attempted in 1966 to foster increased rice production by raising the support price paid for rice by 50%. The support price for corn was raised as well, but not by so large a percentage. Rice production did increase, but rice became more expensive than corn; so more people ate corn and there was a rice "surplus" which was exported. Publicity implied that the Philippines was doing so well agriculturally that there was more than enough food. In fact, the surplus rice was exported at a loss, as the international market price was lower than the subsidy that the government had paid the farmers. The purported surplus existed only because people could not afford to buy the rice at going prices, and not because everyone had enough to eat. Even in locations where the new grains have been unequivocally successful, there are additional factors, such as transportation problems preventing the effective distribution.

Also to be considered is the element of government subsidy as an incentive for farmers to grow the high-yield corn, wheat, and rice strains. Many governments not only subsidize the actual production of the cereals, but contribute to the cost of fertilizers, pesticides, and irrigation. India and Pakistan support wheat production at 100% above the world market price, Turkey at 63% and Mexico at 33%. In fact, the grain quality of these high-yield hybrids is low, so the crops are discounted on the market. These subsidy figures are in reality even higher than one might suppose.

The Risk of Using Hybrid Grains   A potential for agricultural calamity is inherent in the source of the Green Revolution: the hybrid grains. These hybrid varieties lack the genetic variability of the "natural" grains from which they were derived, and are therefore especially subject to the ravages of specific plant pathogens which may infect an entire crop (see Chapter 19). In this respect, the progress made in one area can be considered to be a loss in another (Frankel et al. 1969).

The hybrid grains are prolific, but in other respects they may prove to be inferior to traditional crop plants. Surveys of Philippine farmers who abandoned new rice strains indicate some of the accompanying problems. More than half of those interviewed cited low prices and increased expenses. Another 10% to 15% felt that hybrid rice cultivation entailed too much labor. The market for the high-yield strains is often poor, for cultural and practical reasons (Dalrymple 1969).

As already indicated cultural barriers are at times as effective in impeding the success of innovative agriculture as any other factor. In South China, a local vitamin B deficiency was due to the people's refusal to eat unmilled rice, because they did not like its flavor and the cooking time was longer. Seemingly very minor changes in the nature of a food substance can make it unpalatable. Cultural conditioning in this respect is so strong that people will often reject a new food in favor of more traditional but less nourishing fare. A case in point is that of *Incaparina* (mentioned in Chapter 16), a low-cost, high-protein synthetic flour that has largely failed to satisfy the market for which is was intended.

Economics and the Hunger Crisis   Economics cannot be separated from the question of solving the hunger crisis. Food will not be perpetually distributed free to those in need, either by the

Above In Algeria newly dug wells supply water to irrigate cereal crops. (FAO photo)

Below In Madhya Pradesh, India, villagers deepen a dry pond which will then store more water during seasonal rains. (FAO photo)

affluent nations or the developing countries' own governments. Thus it is imperative that the expense of food be held down as low as possible. This requires low-cost production. Within the agriculturally plentiful United States, 5 or more million individuals are suffering from malnutrition because they cannot afford the high cost of adequate diets. Even if the developing countries were able to grow a sufficient quantity of food for their own people, the cost of the food for an adequate diet must be within everyone's reach. If that qualification is not met, then the goal of adequate nutrition will not be realized, regardless of the quantitative success of agriculture.

Presently, food shortages are eased somewhat by international exchange and donation. The United States and other affluent nations have given large quantities of wheat to India, some African countries, and other nations where food shortages are critical. Substantial amounts have been sold to those that can afford to pay. But eventually, as the world population goes from 4 billion to 7 or 8 billion during the next 30 years, every country that has sufficient grain will need most or all of the crop for its own population. The USSR, which produces twice as much wheat as the United States, purchased a total of more than 11 million tons of cereals during late 1972 and early 1973. Even that amount did not entirely meet the need, although in this instance the concern was with maintaining current levels of consumption rather than warding off starvation. When there is no surplus from wealthier nations to ease famine in less fortunate countries, then the conditions in countries which have not been able to achieve some degree of self-sufficiency will deteriorate rapidly and mass starvation will be inevitable. Today, if serious failings were to arise in crop production, a large percentage of humankind would face starvation. Reserves would last only a couple of years, even in the United States.

Economic factors cannot be divorced from the fate of the Green Revolution. As governmental subsidies are necessary to induce farmers to grow the "miracle" crops, and to cover the expenses incurred by having to use more fertilizers, pesticides, and irrigation, one may reasonably question the degree to which this practise can continue. More than 4 billion dollars per year is given to agricultural subsidies in the United States; yet in this country agriculture represents only about 3% of the gross national product (GNP). In India, 49% of the GNP results from agriculture; in Pakistan, 47%, and in the Philippines, 33%. When a large percentage of a country's GNP is generated by agriculture, then there will not for long be enough incoming capital to support the subsidies.

Above This farmer in
Afghanistan uses the same
kind of wooden plow his
ancestors used 5000 years ago.
(FAO photo)

Left Farm women in Upper
Volta water garden plots.
Learning better ways of
growing food is an important
part of nutrition education.
(FAO photo)

In 1970, on the day that the House Foreign Affairs Sub-committee was conducting a symposium on the Green Revolution, and using the Philippines as an example of a nation that had achieved agricultural self-sufficiency, the Philippine government solicited a 100 million dollar advance from the United States for military bases. The reason for this request was that the Philippine economy was nearly bankrupt. A similar problem faces many small, developing countries. If the use of hybrid, high-yield grains depends on a continuing large capital outlay, the Green Revolution may prove to be impractical—if only from that standpoint. As long as governments are willing and able to offer financial incentives to farmers for cultivating the new grains, then the farmers will not feel that they are subjecting their livelihood to much risk. But if the subsidies become unavailable, there is little likelihood that many farmers in the developing nations, most of whom operate on an extremely marginal profit basis, will spend more cash to cultivate crops which frequently have less market value than traditional ones. Although the yield is greater for the new varieties, the risk is also greater. Farmers existing at near-subsistence levels are, of necessity, conservative. Probably few would be amenable to accepting potential losses from using hybrid cereals if there are no support programs.

As William Paddock (1970) has pointed out, some farmers are probably realizing considerable profits from the Green Revolution. But the most benefits go to those who have the best, and the most land; not to the millions of small farmers struggling for a living throughout the third world. The increased yield per acre is more significant to a farmer with large holdings, whose profits are less dependent on local attitudes toward the grain. From this perspective, the Green Revolution serves more as a boon to agri-business than as a lasting solution for feeding the world's poor. Undoubtedly, yields have increased even for small farmers using the grains. However, as we have seen, the rise in production is based on various elements that the poor are unable to perpetuate if left to rely on their own limited resources.

The problem is a multifaceted one; there are no easy solutions. According to Paul Ehrlich, "Those clowns who are talking of feeding a big population in the year 2000 from make-believe 'green revolutions' . . . should learn some elementary biology, meteorology, agricultural economics and anthropology." Feeding the world's rapidly expanding human population, if this can be done, is not going to be accomplished solely from the development of a few hybrid grains, however high-yielding the crops may be.

The Population Explosion   Even if the Green Revolution could fulfill the aim of feeding all of the world's people at the present time, we would nonetheless be on the brink of crisis in the near future. The essence of the problem is this: if the human population continues its explosive increase, it will soon be impossible to provide enough food for everyone—even if all of the earth's land surface were to be devoted to that end. If increased agricultural production is to lead us toward any future hope, then a worldwide birth control program of some kind will have to be accomplished. Increasing food production in the absence of birth control is analogous to running on a treadmill. Thomas Malthus, in his "Essays On The Principle of Population," offered this observation. "I think I may fairly make two postulata: First, that food is necessary to the existence of man. Second, that the passion between the sexes is necessary, and will remain nearly in its present state. . . ." Somehow we must arrive at a balance between agricultural output and human proliferation.

What are the prospects for a limit to human population growth that can be realized in time to prevent the starvation of millions in the third world? Even such countries as Taiwan and South Korea, where population growth has been slowed, will double their populations within 30 years. In India, when a population control program was begun 17 years ago, the population was increasing at the rate of about 6 million per year. Now, it is growing at an annual rate of 15 million. The actual growth rate has slowed, but the increase is still unmanageable. Consider the plight of Bangladesh, a country about the size of New York State. It is one-sixtieth the area of Brazil, and its population is about 100 million. The entire acreage under cultivation is 21 million, or a fifth acre per person. Nutritional studies indicate that at least an acre per person is needed to provide a minimally adequate diet. Rice is the dietary staple; a very small percentage of the population receives sufficient protein. Although the fertile deltas of the Ganges, Brahmaputra, and some other rivers provide rich agricultural lands watered by monsoon rains, failure of the monsoons causes drought every few years; and when the rains are excessive, extensive flooding may kill thousands and render useless much of the land. Cyclones are also prevalent, occurring at the rate of one or two per year. In 1970, a cyclone and a tidal wave swept over the Ganges-Brahmaputra delta, leaving in its wake an estimated 240 to 400 thousand dead, and millions homeless.

The population of Bangladesh has one of the world's highest growth rates, doubling every 20 years or so. Since much of the potentially tillable land is being farmed, expanding the acreage under cultivation will not be possible for much longer. High-yield hybrids would increase the amount of food that could be grown on the available land. However, in such a case, the Green Revolution is clearly only a stop-gap measure that does not affect the roots of the crisis. If most of the 100 million people of Bangladesh are forced into so marginal an existence at present, how can a solution arise solely from the cultivation of high-yield crops? The best hybrid grains could not keep pace with the population growth rate. Even were that possible for a while, the human population growth rate would eventually surpass the carrying capacity of the land—no matter how productive the crops.

The Green Revolution, then, is not the solution to feeding the world's hungry millions, but only a positive contribution to an exceedingly complex biosocial problem. While the use of hybrid grains has been impressive in some areas, benefits may be transitory, and negative aspects might well be permanent and in some respects biologically devastating (see Chapter 19). For the time being, on a worldwide basis agricultural production has kept somewhat ahead of world population growth. There has been a 3% grain increase in affluent countries which have had a population increase of 1%. In the developing countries, the population increase has been 2.5% against a 3% increase in grain production (Sanderson, 1975). Since 1950, the agricultural output of the world has trebled, while the human population has not. So we should be in a better condition now than previously (Poleman, 1975). Perhaps with improved distribution, as well, the problem may be kept at bay for several more decades.

Ultimately, however, an increasing population in a limited system will grow beyond the resources of that system. The choice is this: there can be a limited population, many of whom can have a quality life, or there can be overpopulation, in which case very few will experience a quality existence. Kenneth Boulding, of the University of Colorado, has added to this concept his "Utterly Dismal Theorem": "If the only check on growth and population is starvation and misery, then any technological improvement will have the ultimate effect of increasing the sum of human misery as it permits a larger proportion to live in precisely the same state of misery and starvation as before the change" (Boulding, 1956). Unless there is a halt to population increase, the Green Revolution cannot be expected to enable humankind to avert famine. More food, but with even more people, does not offer a solution.

Suggested Readings

Borgstrom, George. *Focal Points.* 2nd ed. New York: MacMillan Publishing Company, Inc., 1973.

Boulding, Kenneth. *The Image.* Ann Arbor: University of Michigan Press, 1956.

Dalrymple, Dana. "Technological Change in Agriculture, Effects and Implications for the Developing Nations." Foreign Agriculture Service, U. S. Department of Agriculture, 1969.

Dutt, Ashok K. "Population Pressures In Bangladesh." *Focus* 25 (1974): 1–10.

Frankel, O. H.; Agble, W. K.; Harlan, J. B.; and Bennett, Erna. "Genetics Dangers in the Green Revolution." *Ceres* (FAO) 2, no. 5 (1969): 35–37.

Hardin, Lowell S. Symposium, Subcommittee on National Security Policy and Scientific Developments, House Committee On Foreign Affairs, 1969.

Paddock, William C. "How Green Is the Green Revolution?" *Bioscience* 20, no. 16 (1970): 892–902.

Poleman, Thomas L. "World Food: A Perspective." *Science* 188, no. 4188 (1975): 510–518.

Sanderson, Fred H. "The Great Ford Fumble." *Science* 188, no. 4188 (1975): 503–509.

Tollefson, Bert. Agency for International Development Press, release, 24 February, 1970.

Tydings, Joseph D. (1970). *Born To Starve.* New York: William Morrow and Company, Inc.

CHAPTER

18

MECHANIZED AGRICULTURE: ITS ECOENERGETICS AND ENVIRONMENTAL IMPACTS

On the surface, it might appear that intensive, mechanized agriculture has been an unmitigated benefit. Until quite recently, food production has been so high in the United States that the federal government yearly spent millions of dollars to subsidize farmers for *not* growing certain crops, such as wheat and corn. Fruits and vegetables tend to be larger and more attractive than those grown under less technological conditions, and the per-acre yield of grains has been increased. Mechanization has also reduced the number of people necessary for adequate agricultural output. In recent decades a greater number of individuals from farming families have pursued different lifestyles.

Paradoxically, some of the short-term advantages of mechanized agriculture may be detrimental from a long-range economic and ecological standpoint. In the midst of our abundance, there are subtle factors that call into question the wisdom of continuous agribusiness expansion. Synergistic cycles have been set into motion which, unless these cycles are checked, threaten the future of our most important food crops (Smith et al. 1974).

Economic Considerations of Agribusiness Let us first examine some of the economic aspects supporting the contention that highly technological agriculture is not all to the good. Consider the individual farmers who are attempting to maximize gains. They will be interested in producing the greatest quantity possible on the land available to them. If the crop is a grain, for example, it is unlikely to have any distinguishing qualities that confer an advantage over the crop of other individual producers. Cultivators of fruits and vegetables may focus some efforts on the appearance and size of their crops; but even so, the ultimate quantity remains important.

So a successful farmer will use whatever means are available to maximize the potential yield. It is worth noting that statistical estimates rate only about 25% of U. S. farmers as "successful;" the remainder are marginal producers. Nevertheless, agriculture in the United States more than supplies the needs of the country. Enough food is produced to provide for substantial annual exports as well.

Mechanization. The farmer who can afford to do so will purchase tractors, mechanical harvesters, trucks, and other labor- and time-saving devices, and will invest in chemical fertilizers. Many farmers are dependent on the industries that manufacture these materials; at the same time, the industries are dependent for their survival on agriculture's continuing need for their products. Thus the modern farmer must compete to maximize production, and to maximize production must take advantage of technological innovations. As more machinery is bought, the greater is the financial investment; so that even if an agriculturist significantly increases the yield through mechanization and extensive applications of fertilizers and pesticides, the profit may be modest in terms of inflation. Yet the farmer must continue to make the most of the machinery if the investment is to pay for itself.

Monoculture. An especially disturbing facet of the dependence on high-yield, mechanized agriculture is that it emphasizes monoculture. The costliness of agricultural technology causes the farmer to grow large acreages of the same crop. The serious disadvantages inherent in monocultural practices are, naturally, of less interest to most farmers than the immediate financial necessity.

What are the problems with monoculture? Monoculture simplifies ecosystems. The more simplified an ecosystem becomes, the less stable it will be, and the more vulnerable it will be to disruptions, such as attacks by insects or fungi (see Chapter 19 for further discussion of the biological disadvantages of monoculture). Economically, the farmer who practices monoculture must spend more on fertilizers to replace exhausted soil nutrients, because a single species removes the same elements from the soil year after year, whereas the rotation of several crops can replenish elements used by others. The farmer will have to apply more pesticides, also.

The Effects of Consumer Attitudes. Consumer attitudes play a role in the nature of agribusiness. People usually equate the quality of fruits and vegetables with attractive appearance. Thus, a shopper browsing through the produce section of a supermarket is inclined to look for the largest and most appealingly colored fruits. To compete, a farmer uses all the technological means developed to produce the most attractive produce. No one knows the effects of the increasing residues of toxic materials which we consume in our foods because of the methods used to enhance appearance of produce.

The farmers are caught in a system for which they are not responsible and about which they can do little. Increased productivity per unit of land usually causes prices to drop, because a greater quantity is available. Then the farmer finds it necessary to produce *even more* to maintain the same profit. One cannot blame individuals who are merely attempting to make a living in the context of a complex economic dilemma.

Above *Corn is loaded into a truck from a picker-sheller. Mechanization has been especially applicable to grain cultivation (USDA photo)*

Left *Preparing land for rice planting as a part of the Swamp Rice Improvement Project in Liberia. The side wheels are a special adaptation for working swamp land. Liberia has 700,000 acres of freshwater swamps fed by running streams. (FAO photo)*

However, mechanized agriculture has expanded to the point that most successful farms operate on almost an industrial scale. Many smaller operations have been forced out of business because of an inability to compete adequately with large-scale agribusinesses.

Energy Use

The increased mechanization of agriculture is of particular concern in light of the contemporary scarcity of energy sources. "Modern, high-yield agriculture can reasonably be described as a system that turns calories of fossil fuel into calories of food." (Ehrlich et al. 1973). Farm machinery cannot function without fossil fuels as an energy source. There are also less direct requirements: gasoline powers the trucks that distribute food nationwide; fuel is used in the manufacture of farming machinery. Fuels are consumed in the production of fertilizers and pesticides, both as energy sources and as raw materials.

During the past 50 years, the energy input into food production has increased impressively, not only in terms of energy used for farming but in terms of energy required in the preparation of commercial foods. Even aside from conserving energy, it is relevant that the complexities of food production in developed countries cause costs to be higher, and these expenses are passed on to the consumer. Food in the United States is relatively expensive when compared with most developing nations, but as Americans tend to have a high standard of living, the cost for an adequate diet is not yet prohibitive for the majority. In 1970, an average individual in the U. S. consumed 600 dollars worth of food (Pimentel et al. 1973). Considering that the per capita income is in many countries well below that figure, Americans clearly pay a higher absolute price for their food than do the citizens of the developing nations. Granted, Americans eat much better than do most people in those nations. But if an average East Indian were to pay for his or her food at American prices, it would cost 200 dollars, about twice the annual per capita income for that country.

For every calorie that is eaten by someone in the United States, approximately nine more were required for its production. The energy in the food eaten annually by Americans is about 250×10^{12} food calories (kilocalories) or 1.4% of the total of U. S. energy consumption. Accordingly, about 13% or 14% of all U. S. energy use is in some way involved in food production. The amount of energy devoted to producing food in the United States increased tenfold from 1920 to 1970, whereas the agricultural output was increased by only a factor of two. It is particularly striking that only one-fourth of the energy utilized in the food production system is expended on the farm; processing, distribution, and food preparation consume the remainder. Although the use of energy in food production has continued to grow, the increments of output have been smaller since about 1965.

These figures suggest that, in spite of projections by technological optimists who say that we need only implement mechanized agricultural techniques to allay or eliminate world food shortages, the problem is not that simple. For example, to raise the Indian calorie intake from its present level to 2000 to 3000 per day (the U. S. level) would require more energy if accomplished through mechanization than India currently expends in the total economy. To produce food for the entire world with a U. S. agricultural system would consume about 80% of the world's present estimated energy expenditure. Thus it does not seem likely that the food crisis will be solved by transferring mechanized agriculture to the developing nations. There are not enough energy sources to do so, no matter how the equations are juggled.

An interesting perspective (John and Carol Steinhart, 1974) is that the apparently low number of individuals engaged in food production is misleading. In the U. S. one farmer now feeds 50 or more people, and it is true that fewer farmers are providing more food than ever before. However, "yesterday's farmer is today's canner, tractor mechanic, and fast food carhop." That is, the nature of food production has changed so radically in recent years that many indirect activities related to processing food have come into being. People who previously might have engaged in farming are now employed in food-related industries which previously had considerably fewer jobs.

Increased subsidies of agriculture correlate with the rise in energy use. From 1940 to 1970, total U. S. farm incomes ranged from 4.5 billion dollars annually to 16.5 billion annually, while subsidies increased from 1.5 billion in 1940 to 7.3 billion in 1972. The increased use of machinery has interlocked the role of federal subsidy into our agricultural system. Increased production costs are not balanced by consumer prices, despite the sharp rise in food costs.

Can Energy Consumption Be Reduced?   There are reasonable possibilities for reducing the agricultural energy input in the United States, and such alternatives should be given serious consideration (Pimentel et al. 1973). Better use of manure as fertilizer, instead of chemical synthetics, would reduce pollution from feed lot runoff. The use of manure could save as much as 10^6 kilocalories per acre. There would not be enough manure available to replace completely chemical fertilizers, but its use should reduce energy expenditures. Increased crop rotation and the planting of leguminous cover crops to provide nitrogen fertilizer would also reduce the quantity of commercial fertilizers needed.

The control of weeds and pests could be accomplished in ways more energetically and ecologically conservative. Use of a rotary hoe during cultivation, instead of the application of pesticides, would result in a 10% energy saving, and would reduce pollution as well. Biological pest controls could be more widely used. These techniques include the introduction of sterile males into the pest species' population, the introduction of natural predators, and other "natural" means. Such actions require a small fraction of the energy consumed in pesticide manufacture, distribution, and dissemination, and would lessen the serious environmental pollution that has occurred as a result of excessive and indiscriminate use of toxic chemicals. Also, a more discriminating policy of pesticide use is called for. When feasible, pesticides could be applied by hand instead of being sprayed from airplanes or ground machinery. Although the human labor would be increased, the energy expended in application would drop from 18 thousand to 300 kilocalories per acre.

Plant breeders might concentrate more on qualities of hardiness, resistance to diseases and pests, less moisture content (to reduce the natural gas needed to dry crops), reduced water requirements, and higher protein content—if yields were reduced somewhat by these emphases. Solar energy on the farm has scarcely been investigated. Wind-generated electricity provided by modern windmills is already in use in Australia. And methane gas can be produced from manure or garbage dumps. Such potentials have not been exploited to any degree, and need controlled experimentation and engineering. Hopefully such techniques could supply workable, non-polluting alternatives to the more energy-consuming, environmentally hazardous methods now used in farming.

It has been proposed that chemical farming could be entirely eliminated without disastrous results (Chapman, 1973). Reduced yield per acre would require most land in the soil bank to be put back into production. Some experts estimate that overall output would fall about 5%, and the price of farm products would increase by 10%. Farm income would rise by 25%, reducing or even eliminating the need for governmental subsidies. Consumers are not likely to react favorably to a 10% rise in the price of groceries. However, there would be longterm gains—a reduction in environmental pollution and less energy use, as well as a significant saving of tax dollars.

These might, in the long run, outweigh the immediate displeasing effects. We believe that alternatives to the existing agricultural technology are going to be forced upon us sooner or later, unless new energy sources are developed to replace fossil fuels.

The role and attitude of government is important in changing the agricultural methodology. One aim of the U. S. Department of Agriculture has been to increase agricultural productivity through the development of more chemical agents. The Department finances extensive research to improve and innovate chemical pesticides and fertilizers. It is no small undertaking to shift so broad and accepted an emphasis.

Agriculturists trained in the tradition that increased technology is unequivocally beneficial find it difficult to depart from that view. There is also an economic factor and the vested interests of the corporations that produce the fertilizers and pesticides. However, a change is imperative if we are to achieve a rational agricultural system not dependent on a high and ever-increasing energy expenditure. Surely long-range interests would be better served by research into some of the biologically conservative possibilities suggested in the preceding paragraphs.

Soil Conservation   A problem that is not unique to mechanized farming, but which may, in some cases, be aggravated by it is soil erosion. Repeated plowing during the growth of crops causes a continuous and cumulative loss of topsoil. It takes thousands of years to build up a single inch of topsoil, where *inches* may be lost in a single year. The practise of using the rotary hoe instead of herbicides, and planting cover crops would alleviate soil loss to some degree, although the cover crops might have the disadvantage of competing for moisture and nutrients.

Topsoil conservation is important to future agricultural welfare. It is an issue that must be dealt with in any agricultural system, not just technological ones. The environmentalist Aldo Starker Leopold related a dismal tale of how quickly human beings can ruin the productivity of the soil (Leopold 1957). Less than 200 years ago the foothills north of Mexico City were covered with forests of pine and oak. As the population of the Valley of Mexico grew, more land was cleared and planted in corn and wheat. Within a few years, the topsoil had been washed away and minerals had been leached out—leaving a barren subsoil that would no longer support corn and wheat crops. However, it was sufficient for the hardy *maguey* plants, from which fibers and tequila are obtained. Eventually, even the subsoil eroded. Under the subsoil there was a layer of hardpan on which not even maguey could be cultivated. Still, a few desert plants grew there, providing food for limited herds of goats and donkeys which soon eliminated even this sparse vegetation. Today, the area is totally a wasteland, productive of nothing.

The story is not an isolated one. As we have previously mentioned, the Sahara desert of Africa and the Thar desert of western India were lush only a few thousand years ago, and were altered drastically, largely due to deforestation and poor agricultural practises. To prevent the loss of topsoil, which is, for all practical purposes, irreplaceable, technological farming techniques should be balanced by practises that will minimize soil loss. Another "dust bowl" is not an impossibility, even in the United States.

It is clear that modern agriculture has given us more food per acre, but the effects of mechanization must also be measured in terms of energy consumption, environmental degradation, economic factors, and the biological hazard of monoculture. The current trend in agricultural energy input, to focus on continuous production increase, cannot continue unabated. For example, 10^3 kilocalories of fuel are required to cultivate a square meter of corn. At present, the yield from that land gives us 2×10^3 kilocalories of food. Presumably, more energy will be expended in the future, with little probability of further increases in yield. Clearly a more conservative, stable, and balanced system of agriculture must be developed—a more self-contained system less dependent on complex machinery and chemicals.

Suggested Readings

Chapman, Duane. "An End to Chemical Farming?" Environment 15, no. 2 (1973): 12–17.

Dasmann, Raymond F. Environmental Conservation. New York: John Wiley & Sons, Inc., 1972.

Ehrlich, Paul R.; Ehrlich, Anne H.; and Holdren, John P. Human Ecology. San Francisco: W. H. Freeman and Company, 1973. "Fuel and Food." Scientific American 230 (1974): 48.

Hardy, R. W. F., and Havelka, U. D. "Nitrogen-Fixation Research: A Key to World Food?" *Science* 188 (1975): 633–643.

Hirst, Eric. "Food-Related Energy Requirements." *Science* 184 (1974): 134–138.

Leopold, Aldo Starker. *Wildlife of Mexico.* Berkeley: University of California Press, 1959.

Pimentel, D.; Hurd, L. E.; Bellotti, A. C.; Forster, M. J.; Oka, I. N.; Scholes, O. D.; and Whitman, R. J. "Food Production and the Energy Crisis." *Science* 182 (1973): 443–449. *Science,* Volume 188, no. 188. May 9, 1975.

Smith, J. C.; Steck, Henry J.; and Surette, Gerald. *Our Ecological Crisis.* New York: Macmillan Publishing Company, 1974.

Steinhart, John S., and Steinhart, Carol E. "Energy Use in the U. S. Food System." *Science* 184 (1974): 307–316.

CHAPTER

19

MONOCULTURE AND THE GENETIC FUTURE OF MAJOR CROPS

The emphasis of the Green Revolution on the high-yield hybrid crop plants may, over time, prove a threat to our food supply. Hybrids show only a limited expression of the total genetic variability characteristic of wild species. Dr. Jack R. Harlan, a professor of plant genetics at the University of Illinois, commented in a panel discussion sponsored by the United Nations' Food and Agricultural Organization (FAO) that "the food supply for the human race is seriously threatened by any loss of variability." (Frankel et al. 1969).

Many of the qualities selected for in plant hybridization have been for the producer's benefit. Whether or not the characteristics benefit the consumer is a consideration that is frequently secondary. Of course, there are exceptions. Norman Borlaug developed high-yield wheat for the specific purpose of helping to alleviate worldwide hunger, and a great deal of other research has been conducted in efforts to provide more food for a growing human population. Many characteristics of crop plants are subject to manipulation, however, and quality has suffered in some cases. Crop yield, resistance to disease and to insects, ease of mechanical harvesting, color, flavor, and texture are among the properties of crop plants that are subject to genetic engineering.

Loss of Genetic Variability By far the most detrimental aspect of hybridization is the loss of genetic variability. Widespread propagation of high-yield hybrids has caused increasing acreage to be planted in the same crop, whereas formerly there was greater diversity. Throughout southern Asia, for example, there has been extensive planting of a few wheat varieties that produce high yields, and all exhibit a high resistance to the same kind of wheat rust. The danger is that if a new variety of wheat rust were to appear—millions of acres of crops could be devastated in a very short time. The genetic variability responsible for producing resistant individuals may have been eliminated in the hybrid strains.

The possibility of such a disaster is not just speculative. In 1946, some 30 million acres (about two-thirds of the crop) of U. S. farmland were planted in a hybrid oat variety resistant to the "Victoria type" rust. A new rust disease, unknown four years earlier, appeared and attacked the crops so effectively that by 1948 the new hybrid had almost vanished.

In the early 1950's, the control of a wheat fungus disease known as "stem rust" was a highly developed aspect of plant science. A certain strain of the rust (15B) had been under observation for 10 years. Yet, during 1953 and 1954 an epidemic of that strain nearly wiped out the entire U. S. crop of durum wheat. Consider a similar hypothetical circumstance in India, where there are too few technicians to monitor the diseases developing in wheat and rice fields. With increasing dependence on a few hybrid varieties of wheat and rice, the effects of a new disease could be devastating, ruining a large percentage of the crops. In a country where food production cannot keep up with population growth under normal circumstances, such a crop failure would have tragic results.

Monoculture has been responsible historically for initiating food crises. Ireland's potato blight of the 1840's resulted in the deaths of 2 million people, and caused the emigration of numerous others. Three million bushels of wheat in the United States and Canada were destroyed in 1916 by red rust; and in 1946, when a virulent fungus attacked rice in Bengal, many thousands died. As mentioned previously, during that same year millions of acres of oats were ruined by rust.

It is most unlikely that technologists will be able to develop hybrid or non-hybrid crops that are resistant to all the potential diseases or insect depradations that may occur. High-yield rice (IR-8) has been found to be susceptible to a virus carried by green leafhoppers, and is also vulnerable to the insects themselves. The problems inherent in concentrating on a few hybrid crops can, in some respects, be minimized. A greater variety of improved strains should be engineered to allow a larger number to be cultivated within a given area. However, this will be possible only if we retain the raw materials—*the genetically varied, primitive stock*—from which breeders will be able to produce new strains. Such action might seem relatively easy to effect; but, in fact, the wild relatives of our crop plants are vanishing at a disconcerting rate.

The Importance of Wild Strains   John Creech, of the USDA's Agricultural Research Service (ARS), points out that plants have always been low in the order of genetic resource priorities. "The attitude toward seed conservation has been one of reaction; if we need rare strains to breed a stronger variety of grain in the event of an epidemic, we go and collect it. We have felt that we can always go to the country of origin and get them" (Miller 1973). However, expeditions in search of native seeds are discovering more and more frequently that the wild varieties are no longer to be found. The disappearance of these plants may be related to the use of hybrid strains and the destruction of natural habitats as a result of the "development" of wilderness areas. As nature can no longer serve as a permanent reservoir for genetic variability, it is surprising that so little effort has been made to establish seed banks. The loss of genetic variability is *not* a reversible phenomenon that can be

Wheat rust (Claviceps purpurea) on wheat. (USDA photo)

corrected as an afterthought; ". . . once this germ plasm is lost, it is lost forever. There is no way in which it can be recovered. Some have felt that one could generate variability by radiation or chemical mutation of genes and so forth, but, in practise, this has not been the case" (Harlan, in Frankel et al. 1969).

Genetic Conservation Interest in conserving the genetic resources of plants has been rather weak in the United States—especially in view of the relative scarcity of native food-producing species. In spite of the success of U. S. food production, most of the wild grain varieties, on which we depend for genetic variability, have been imported or hybridized from imported varieties. As described in the preceding chapter, the productivity of the U. S. agricultural system is based on monoculture. This being the case, the crops are vulnerable to disease and insect epidemics.

The USDA's Agricultural Research Service has the responsibility of finding, storing, and perpetuating a reservoir of wild species. When the USDA was established more than a hundred years ago, plant introduction was a high priority. Over the decades, more than 400,000 kinds of plants and seeds have been brought into the country as potential food plants. Today, the ARS maintains the National Seed Storage Laboratory in Fort Collins, Colorado and four regional plant centers on an annual operating budget of less than a million dollars. In the context of rising costs, this amount is scarcely enough to maintain the existing program, and leaves no room for a much-needed expansion of facilities and operations.

A report issued in 1972 by the National Research Council of the National Academy of Sciences stated that "most major crops are impressively uniform and impressively vulnerable." A second report in 1973, produced by a USDA subcommittee on plant genetic problems and the National Association of State Universities and Land Grant Colleges, goes further to suggest that the current trend toward the disappearance of resources for variability is "dangerous and deteriorating rapidly." The report cites efforts to manage these resources thus far as being "haphazard, unsystematic, and uncoordinated."

The world's botanical pool of genetic diversity has been drained in a relatively brief period, so brief perhaps that many individuals engaged in agricultural research have not yet appreciated the potential severity of the impact. Less than a hundred years ago, most cultivated plant varieties in any part of the world were "primitive" ones. This is still the case in some of the developing nations. However, as the use of hybrids and the practice of monoculture spreads, natural diversity will continue to lessen unless very emphatic steps are taken to prevent it.

In Africa, for example, there is a native rice that grows in the highlands. Due to a relatively low yield, it is being replaced by high-yield hybrid rice strains. Yet the native plant evidences a strong resistance to drought, a quality that could prove immensely useful in dry areas where irrigation is too expensive or is for other reasons not feasible. If research with this variety is delayed much longer, the native rice may cease to exist and the basis for a drought-resistant rice will have been lost. The immediate advantages of high-yield Asian rice may cause the elimination of valuable genetic characteristics of the native rice that could be employed against famine during dry years. Whether or not this African rice is developed into a useful crop before it is extinct is an example of a larger problem: there will not be time to investigate the genetic possibilities of wild food plants that are now disappearing unless the plants are conserved in the present. It is doubtful if all potential variability could be retained, even with assiduous conservation efforts. However, enough would be maintained to provide breeders with sufficient raw material, which is what matters from the standpoint of food production. It would be impossible to utilize all existing natural variations.

Organizational Efforts Toward Conservation   The Food and Agricultural Organization of the United Nations has been the foremost promoter of genetic conservation. In February 1973, O. H. Frankel issued the first worldwide itemized survey of extant genetic resources, identifying those that are menaced with the possibility of extinction. The USSR, while not a member of the FAO, has been sponsoring a partnership with that organization a school to train individuals from the developing countries in plant exploration. Also, the Rockefeller Foundation has recently taken an increased interest in this field, having considerably enlarged the budget for genetic conservation, and establishing committees to study the loss of genetic variation in wheat, rice, sorghum, and maize. The Rockefeller Foundation support has been contributed primarily through the Consultative Group on International Agricultural Research, a collaboration of 29 nations, foundations, and agencies committed to the long-term financial support of international agricultural research. This latter group represents perhaps the most hopeful action to date in organizing efforts to reduce genetic erosion. Even with greater awareness and a resulting increase in positive action, a monumental effort will be required to salvage a broad spectrum of genetic variation for the future.

International cooperation is a necessity if these efforts are to develop into a successful outcome. Seed storage should be implemented on a national and an international basis, with a free exchange of germ plasm. Inevitably, the developed countries will carry much of the financial cost. The storage of viable seeds is considered the best means for maintaining stocks and reducing the possibility of changes in desired characteristics. Seeds would have to be subjected periodically to germination tests, and cataloguing systems would be necessary to provide easy access to the variety of plant material in storage. Without adequate budgets, such storage centers will not be established.

Sources of Genetic Variation   For any given plant, the best source of genetic variation is usually found at the place of origin, called a *primary center*. It was once believed that diversity spread exclusively from centers of origin; it is now known that this is not necessarily the case. Extensive diversity also may be found in secondary centers far from the site of origin. In the context of conserving genetic resources the area of origin is essentially irrelevant. What is of significance is whether or not the variation exists. For this reason, genetic resources must be sought not only at the known point of dispersal for a species, but in any areas where wild individuals occur.

Species distribution is far from uniform. As with animals, the number of species present is related to the latitude, rainfall, humidity, and other factors. Costa Rica and El Salvador, for example, host as many indigenous species as do the United States and Canada, although the former constitute 1% of the land area of the latter. Countries having a marked diversity of habitats will yield more varieties of plants, because different environmental conditions will have effected a greater range of responses during the plants' adaptive evolution.

At present much of the developing world, where nutritional problems are the most severe, is still rich in food plant variation. In this respect, these countries are protected from the potential disasters that may accompany extensive monoculture. Yet it is clear that this kind of safeguard can be dissipated quite rapidly by the cultivation of hybrid strains. The problem is not insoluble. The effective collection and storage of genetically varied resources should be an inseparable component in the development of high-yield hybrids. If this is not done, then in the future we may well be confronted with agricultural crises of heretofore unknown dimensions.

Locating and conserving varied plant strains is not as simple as one might suppose. It requires a great deal of field work, which consumes both time and money. Valuable resources can be lost over a short period in relatively small areas. Even with intensive programs to preserve genetic variability, we must accept the fact that it will be possible to salvage only a small percentage of the strains which now exist. When the Aswan Dam was built in Egypt, seed strains were irretrievably lost. Without strict care to prevent such accidents, even the seeds in storage banks could be lost; unless the same varieties are not distributed to at least several centers. An extensive collection of millet seeds was discarded in central Africa upon the owner's death. The risks would be less in official storage areas, but even so, losses from fire, theft, and carelessness are more than a remote possibility.

Hybrid crops can continue to contribute to efforts against hunger, but certain stipulations should be regarded. The hybrid plants should be developed and planted in sufficiently diverse areas within any locality so that there will be less danger from attack by a disease or other pest. More research devoted to developing the potential from local plants, such as the African drought-resistant rice, could diminish the likelihood of monocultural crises. In the near future, the most pressing need is the conservation of primitive, genetically-varied plant stocks. The loss of these stocks would place all humankind in a precarious position—possibly endangering the majority of our plant food resources.

Suggested Readings

Frankel, O. H.; Agble, W. K.; Harlan, J. B.; and Bennett, Erna. "Genetic Dangers In the Green Revolution." Ceres (FAO) 2, no. 5 (1969): 35–37.

Harlan, Jack R. "Our Vanishing Genetic Resources." Science 188, no. 4188 (1975): 618–621. Science, Volume 188, no. 4188. May 9, 1975.

Miller, Judith. "Agriculture: FDA Seeks to Regulate Genetic Manipulation of Food Crops." Science 185, no. 4147 (1974): 240–242.

Miller, Judith. "Genetic Erosion: Crop Plants Threatened by Government Neglect." Science 182, no. 4118 (1973): 1231–1233.

Paddock, William C. "How Green is the Green Revolution?" BioScience 20, no. 16 (1970): 897–902.

CHAPTER

20

MARICULTURE: SEAWEEDS, PHYTOPLANKTON, AND THE MYTH OF FOOD FROM THE SEA

Characteristics of Algae   The term "algae" refers to several members of the plant kingdom that evidence diverse modes of growth, reproduction, and metabolism. Algae can be found in a range of habitats. Some are benthic (bottom dwelling) in marine and freshwater environs. Some are planktonic and can be found floating or drifting in the ocean or in fresh water. Seventy percent of the oxygen in the world's biosphere is produced by these algae. Some are terrestrial and are often important components of soils. Some algae, *cryophilic,* actually thrive in ice and snow; and *thermophilic* algae are a major botanic component of hot springs.

Many algae are colorless and nonphotosynthetic, qualities which justify the inclusion of algae with the *Protista.* However, most algae are characterized by the following features: the presence of chlorophyll; a rigid cellulose wall that may include carbonates or silicates; a lack of protective cells surrounding developing reproductive cells; and an absence of nutritive provision to aid the developing zygote. In size algae range from the smallest *eukaryotic* cell, which is approximately 1μ in diameter, to sea kelps, which may attain lengths of more than 35 meters. As phytoplankton, algae form the base of aquatic food chains in both freshwater and marine ecosystems. Algae store a variety of products, from starches to oils. Fossil oil deposits may be largely the food reserves of ancient algae.

Algae are not particularly conspicuous in terrestrial environments and are often unnoticed in freshwater ecosystems. It is in the underwater habitats of inshore marine environments that algae are most apparent. From such habitats come most of the algae that are directly consumed or used in other ways by human beings.

Economic Utilization of Algae.   Economic uses are many. Algae are consumed directly as food and the cell wall polysaccharides (sugar polymers) used in a variety of products from cosmetics to paints to ice creams. Algae are given as fodder to livestock in some coastal areas and are employed as livestock food additives. The ash from algae is a component of some high-quality soaps and glass. Algae are used in organic fertilizers, and in a variety of microbiological assays. In such ways algae have a small but not insignificant place in the economies of many countries.

Nutritional Value.   The nutritional value of algae is primarily, but not exclusively, as a source of trace metals and vitamins. Trace metals frequently constitute as much as 20% of the dry weight of algae, which are also rich in vitamins A, C, E, and some members of the B complex. Algae vary in protein content, ranging from 1 to 40%, averaging 5% to 10%. It might seem, from these figures, that the value of algae would be as a protein source. Much of the protein, however, is bound to the cell wall and chemically constructed in such a way that human enzyme systems cannot break it down. Thus, much of the algae protein is unavailable. However, significant amounts of protein can be assimilated from the algae containing very high protein percentages.

Since early historical times, algae have been used as vermifuges.* Benefit in feeding to livestock is due to weight gains resulting from lessened incidence of parasites in the animals. Some algae also have rather strong antibiotic properties, contributing to lower infection rates in animals, and consequent weight gains. As an ethnic dietary item, algae have a long-standing tradition in Polynesia, China, Japan, the Philippines, Indonesia, and some countries of Northern Europe. Thus algae can be a source of vitamins, trace metals, and sometimes antiparasitic agents.

Algae Nutrients

Agar. The cell wall polysaccharides of algae are commercially important. Agar, carrageenin, and alginic acid are derived from the cell wall. *Agar* is a polysaccharide that is water extracted from red algae (*Rhodophyta*). The product has especially valuable properties: it gels at 0.5% to 1.0% concentrations; it remains a liquid at temperatures as low as $-59°C$, and liquefies at 90°C. Toward the end of the 19th century, agar was discovered to be a valuable aid in the development of the science of clinical microbiology. Prior to that time, gelatin from the matrix of animal bones had been the only solid agent available as a medium for microbial growth. Many bacteria have the capacity to liquefy gelatin, whereas few microorganisms are able to liquefy agar.

To prepare agar, red algae are collected and the water content of the algae extracted under pressure. The extract is clarified, filtered, and stored at a cold temperature for two days; then the water-soluble impurities are removed. The product is further purified and bleached translucent. Most of the agar used in medical and pure scientific research comes from the red alga genera *Chondrus* and *Gelidium* that grow along the northeastern Atlantic Coast of the United States and the coasts of Indonesia. In New Brunswick and Nova Scotia, *Chondrus* is collected from boats or at low tide with long-handled rakes; then it is piled in heaps on the shore, partially

Serving to expel worms or other parasites from the intestinal tract.

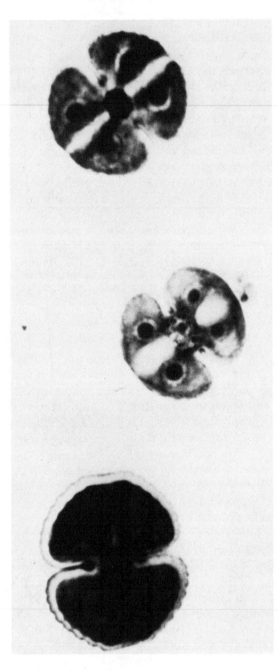

Above *Freshwater unicellular phytoplankton. (Photo by the authors)*

Right *Gelidium, a genus of red algae from which agar is extracted. (Photo by the authors)*

dried, and sacked for shipping to processing plants. In 1966, more than 300 tons were harvested in Canada, at a market value of 50 thousand dollars. During the same year, 16 thousand metric tons of agar were processed from algae collected worldwide in Japan, Korea, Canada, Indonesia, the Philippines, and South Africa.

Carageenin. Another important polysaccharide, carageenin is also extracted from Chondrus. Like agar, it is a complex polymer composed of several linkage groups. It, too, is obtained by water extraction, after which potassium is added to increase the gelling properties of the molecules. The final product is used as a stabilizer and emulsifier in commercial eggnog and chocolate milk, and also to create a colloidal suspension of yeast during the preparation of beer.

Alginic Acids. These are derived from brown algae (Phaeophyta) such as Laminaria, a common seaweed of the North Atlantic, and Macrocystis, a giant kelp found along the Pacific Coast from Baja California north. The acids comprise 14% to 40% of the cell wall material of these plants. The molecular structure is similar to that of the pectic acid (pectin) in the fruits of higher plants. Alginates also serve as hydrophilic colloidal agents, and are used as thickeners, cosmetic additives, printing pastes for textiles (an ancient Oriental use), latex additives, glazes for papers and ceramics, and emulsifiers, such as for water insoluble antibiotics—an example is aureomycin.

Other Uses and Effects **Fodder and Fertilizer.** Traditional uses of algae in northern Europe are as a crop fertilizer and an animal fodder. With some technological modifications, algae is still used as animal fodder. In Scandinavia, considerable research has been devoted to combining algal additives with livestock feed. The practice prevents wool shedding in sheep, reduces the incidence of live parasites in hogs (with accompanying weight gains), increases butterfat content in cow's milk, and increases iodine content and yolk quality in the eggs of domestic fowl. The feeding of algae to livestock would probably be beneficial in any area where animal diets are marginal. The algae supplies vitamins and minerals as well as conferring the above advantages.

Seaweeds are rich fertilizers because of the significant quantities of potassium, phosphorus, and trace elements. The plant material may be spread out to rot in fields or mixed in with other organic fertilizers. The algae enhance the nutrient content of the soil, and also aid in soil binding and in the breaking down of clays and other hard soils. It has been shown that blue-green algae, if allowed to flourish in water-logged fields, will lower the pH and increase the content of nitrogen, phosphorus, and other organic matter; thus markedly improving the fertility of the soil. Some algae are also agents of nitrogen fixation. The result of adding the blue-green alga *Tolypothrix* to rice paddy fields was to increase the crop 2% the first year and 8%, 15.1%, 19.5%, and 10.5% respectively for subsequent years. (Watanabe, 1962, in Zajic, 1970).

Seaweed Meal. More than 50 thousand tons of seaweed meal are produced annually in Canada, France, Great Britain, Ireland, Norway, South Africa, and the United States. Most of the meal is obtained from the brown alga *Ascophyllum nodosum*. The meal can be a valuable supplement in human diets, especially in areas where iodine is a limiting factor, as all seaweeds have a high iodine content. As a source of iodine, algae is a preventive for goiter.

Other Commercial Applications. Another commercial application takes advantage of the remains of long dead algae. During the Tertiary and Quaternary periods, large blooms of *diatoms* (planktonic algae of the division *Bacillariophyta*) prospered. These remain as deposits of diatomaceous earth, in such locales as Lompoc, California. Such deposits are the remains of the silicified cell walls of those unicellular algae, and are about 88% silica in composition. Known commercially as Fuller's earth, the deposit is used as a toothpaste abrasive, a filler in paints, and as a filtration material in the preparation of antibiotics and in wine production. About 300,000 tons of Fuller's earth are used annually.

In scientific application, algae are employed as bioassay organisms. The pharmaceutical production and quality control of Vitamin B_{12} is an example of this use. Certain flagellate algae require this vitamin in precise amounts for growth, so the quality of the vitamin may be assessed by the nature of algae growth in the presence of unknown Vitamin B_{12} concentrations.

Sodium laminarin sulfate, a blood anticoagulant qualitatively similar to heparin, is a product of the brown alga *Laminaria*. Also, some algae manifest strong antibiotic effects against microorganisms. The diatom *Nitzschia palea* is used to treat public water supplies, being grown in sand filtration beds to inhibit the proliferation of *Escherichia coli*, the common bacterium of the human colon.

Damaging Effects. Certain algae may also have considerable negative economic significance. Some blue-green algae (*Cyanophyta*) and dinoflagellates (*Pyrrophyta*) produce toxic substances which may become concentrated in fish and aquatic invertebrates such as clams, mussels, and oysters—rendering them unfit for human consumption. The introduction of too great a quantity of nitrates and phosphates into water systems can result in *eutrophication,* a process wherein the explosive blooming of algae (of a number of genera) results in the depletion of oxygen and a subsequent decline of biological diversity in aquatic ecosystems. Ultimately, eutrophication means the "death" of the ecosystem.

Seaweed as Food   Algae have received the most attention, both in ancient and modern times, for potential value as human food. A number of genera, including the brown alga *Laminaria*, the red alga *Porphyra*, and the green alga *Monostroma*, have been cultivated in sophisticated mariculture systems for thousands of years. In Japan, the cultivation of *nori* (*Porphyra spp.*) is of major importance in mariculture. The annual harvest is about 200 tons. Its current importance is due in large part to the 1950's research of the late K. M. Drew-Baker. She demonstrated that an alga, known at the time as *Conchocelis*, found growing on the insides of oyster shells is in fact an alternate phase in the life history of *Porphyra*. Her discovery of this fact enabled the Japanese to develop techniques for artificial spore-cultivation. These techniques are of major economic significance today. In *nori* production, a spore-producing stage of the plant is cultivated in cement tanks. The released spores are allowed to settle and attach themselves to substrates such as bamboo, coarse netting, or poles which are later transferred to shallow, protected bays. There, the plants mature and, after three months of growing, are harvested. After harvesting the algae are macerated and poured onto frames to form a felt of fragments in a process identical with paper manufacture. After drying, the sheets of seaweed are commercially processed.

   Nori is remarkable for its high protein content (up to 30% at dry weight), its high content of vitamins A, B, C, and E, and an abundant supply of trace metals. It is eaten as a condiment with noodles, as a wrapping for vinegar-rice sandwiches (*sushi*), and as an ingredient of soups.

   Kombu refers to brown algae (*Laminaria spp.*) harvested by itinerant workers who travel along Japan's coasts collecting the seaweed when it is ready for gathering. It is dried, commercially graded, and marketed as a stock flavoring for traditional soups. The exotic bird's nest soup of China owes its characteristic flavor to a blue-green alga growing in the "nest."

Future Potential In recent years, much has been written about the potential of algae as a protein supplement to aid in feeding the world's expanding population. Several green (*Chlorella spp.*) and blue-green (*Spirulina spp.*) algae have been cultivated for this purpose. The problems involved in producing unicellular algae as food center mainly on the difficulties in cultivating sufficient quantities and in neutralizing the rather strong flavors to make the algae acceptable to conventional palates.

In spite of the economic importance of *nori*, it is inferior as a food to *Chorella* algae, which grows more rapidly and under a greater variety of conditions. *Chorella* has a higher content of proteins and lipids than does *Porphyra*, and is richer in vitamins. Nutritional quality, simple requirements for cultivation, the production of minimal waste material, and rapid growth make it an easy crop. It is currently cultivated in Germany, Japan, and the United States.

Microalgae such as *Chlorella* and also *Spirulina* are quite high in protein, about 35% to 40%. The only land plant offering significant competition in this respect is the soybean, which is approximately 35% protein. The nutritional content of *Chlorella* is similar to a mixture of soybeans and spinach. Under existing economic conditions, however, the algae fail to compete with the land crops. The production of *Chlorella* protein costs approximately 50 cents per pound because stringent technological controls must be exercised to keep the system productive. In contrast, soybean protein costs about 6 cents per pound, a difference great enough to suggest that it may be a long while before we sit down to meals of *Chlorella*. Another factor is the flavors of seaweeds which are not particularly esteemed even in countries with a lengthy tradition of seaweed dietary use. Unless, for some reason, land priorities are drastically shifted, terrestrial production of protein will continue to be the most viable approach to meeting food needs.

Seaweed Cultivation Nonetheless, research in seawood cultivation is likely to increase in the pursuit of more productive and varied food resources. In United States universities with marine biology programs, seaweed maricultural research has been actively engaged in. For example, the Departments of Botany and Oceanography at the University of Washington are involved in a joint long-term venture to investigate possibilities for developing the cultivation of marine seaweeds (Jamison and

Beswick, in Nisizawa 1972). Red seaweed is currently harvested from wild beds by the Lummi Indians, but these natural sources are limited, and represent an economic yield of at most half a million dollars. But since more than two million acres of submerged land is part of the state of Washington, it is assumed that the harvest of seaweed could be vastly increased through cultivation. Laboratory experimentation on the environment for seaweed cultivation have been combined with a field program studying the effects of transplanting *Iridaea* and other species.

In this particular case, it appears that the nature of the substrate in Puget Sound is the primary limiting factor. If so, the expansion of a seaweed industry in Washington (and probably in many other areas) will depend on altering the substrate to make it more favorable to the desired varieties of seaweed. From an economic standpoint, the commercial increase in production of seaweeds in the United States will require mechanical harvesting due to the high cost of manual labor in this country.

There are also many difficulties intrinsic to the seaweeds (or algae). In Japan, the cultivation of *Laminaria* from juvenile plants produced by spore collection resulted in plants inferior to naturally growing ones. This was the case for two years, until a new technique was developed. Growing the spores in a special medium in rooms with a fixed temperature yielded plants that were, in one year, equal in quality to two-year-old natural ones (Hasegawa, in Nisizawa, ·1972). Furthermore, analysis of the upper region of the blade of these "forced" seaweeds revealed high contents of protein, nitrogen, glutamic acid, and proline. Under experimental observation, a yield valued at 720 dollars was obtained from a total investment of ·160 dollars, demonstrating that cultivators can realize a substantial profit from using starter plants begun with forced cultivation techniques.

Algal Resources Because marine environments are so great and varied a part of the earth's surface, it is not easy to measure the existing resources of marine algae. Currently, the Food and Agricultural Organization (FAO) of the United Nations is attempting to prepare a summary of seaweed resources, based for statistical purposes on major fishing areas established to provide world fishery statistics. These figures are intended to include estimates of harvestable quantities of seaweeds and the amounts actually being harvested.

Preliminary figures suggest that algal resources globally are being reduced. In 40 years, the width of seaweed belts along rocky coastal bottoms was reduced 50% in water considered to be unpolluted. Plants were found growing at levels only half as deep as those previously recorded. A major factor in this marked reduction may be an increase in surface phytoplankton, resulting in less light penetration. Despite the assumption that the waters in question were unpolluted, the increase of phytoplankton is probably due to greater outfall of nutrients into coastal waters. At present, there is little quantitative information available on algal growth, so it is extremely difficult to form an accurate assessment of current trends. The FAO's census is a move in the right direction, and will provide a basis for comparison in the future. Seaweeds are at present an "underused" resource. As such, they have not been subject to the same scrutinies as marine fishing.

There are differences, however. Fish and shellfish seem to have an intrinsically more appealing flavor than do seaweeds. The nutritional value is considerably higher. Seaweeds are more expensive to produce, and it is doubtful if even good marketing techniques could enable them to become a popular dietary constituent when there are so many alternatives. At present, in our opinion, algae culture cannot be looked upon as a potential solution for global food shortages. The possibility is hindered by cost, cultural reservations, and logistic difficulties in cultivating seaweeds in quantity.

Suggested Readings

Jackson, Daniel F., ed. *Algae, Man and the Environment*. New York: Syracuse University Press, 1967.

Kumar, H. D., and Singh, H. N. *A Textbook on Algae*. New Delhi, India: The Affiliated East-West Press, Ltd., 1971.

Nisizawa, Kazutosi, ed. *Proceedings of the Seventh International Seaweed Symposium*. Tokyo: University of Tokyo Press, 1972.

Zajic, J. E., ed. *Properties and Products of Algae*. New York: Plenum Press, 1970.

CHAPTER

21

INNOVATION IN MEETING WORLD FOOD DEMANDS

Innovative thinking will be a valuable asset in increasing the world's usable food supply. The Food and Agricultural Organization of the United Nations has all too often balked at supporting nontraditional measures for meeting the food crisis. New agricultural approaches will be of value, but we must also exploit other potential food sources. There is not necessarily going to be a solution for much of the world's peoples until population growth levels off; yet ethical motivations will spur research to find new food sources among a range of relatively untried possibilities.

Prevention of Crop Loss   In agriculture, new techniques are being employed to prevent crop losses, perhaps the most basic avenue for increasing food supplies (see Wittwer, 1970). It is estimated that more than a total of 75 million acres of crops are sacrificed annually in the United States alone to insects, diseases, and weeds. If such depredations could be reduced through natural controls, instead of by environmentally-damaging chemicals, the long-range results would be safer and more effective. It is now possible to synthesize the juvenile hormone of insects, to which most species respond. As little as one gram per acre dispersed over fields will keep insect larvae from pupating or maturing. The unprotected larvae are then killed by winter in northern climates; and even if death does not result, the larvae will not be capable of producing new generations.

Increasing Yields   There is the possibility of increasing the agricultural potential of sandy, xeric soils. Thin asphalt strips about two feet below the surface create a water seal. In control situations, this method has increased yields of tomatoes, cabbages, beans, and other crops by 60% to 80%. Not only is the amount of water required reduced considerably, but soil nutrients are less leached out.

Raising atmospheric levels of carbon dioxide (CO_2) will usually result in notable yield increases if available CO_2 is a limiting factor. The experimental infusion of CO_2 into greenhouses has raised production of tomatoes, cucumbers, and lettuce by 25% to 50%. Other basic crops of more economic and nutritional importance, such as corn, grain sorghum, soybeans, barley, and rice have also given higher yields when CO_2 levels were elevated. The use of CO_2 fertilizers also results in a greater total crop.

Productivity has been increased in arid lands by using plastic, low-cost greenhouses that maintain high humidity and high CO_2 levels. A research project on the Gulf of California near Puerto Peñasco in Sonora, Mexico has demonstrated that crops may be grown in this way in an environment that is intrinsically hostile. In this experimental location, there is an abundance of sunlight, but there are essentially no sources of fresh water, a factor creating a most formidable aspect for both people and crops. In these greenhouses, the CO_2 content is maintained at a concentration of 1000 ppm (parts per million), and humidity is 100%; so watering requirements are 1% to 5% of what would be normal in nature.

Dietary Innovation Although the use of biological controls and innovative means in inherently unreceptive environments will increase food production, innovation will tend to be limited to the *kinds* of foods we eat as well as to the *ways* in which foods are produced. As we have previously indicated, dietary expectations tend to be culture-bound. The introduction of alien fare is likely to meet with resistance. However, dietary can be accomplished, especially if attempts are made to integrate new foods with those that are already preferred in a given area.

One theoretical solution is to culture single-celled organisms on petroleum and other substrates. Such single-cell protein (SCP) could be produced in quantity; and, if it can be prepared economically for human consumption, it might help alleviate a significant amount of the world protein deficit. Aside from building production plants and arranging for distribution, a major problem in the use of SCP may be to persuade people that it is a palatable food. Also, since petroleum is often designated as the most favorable growth medium, legitimate reservations may be raised in regard to the scarcity and expense of fossil fuel supplies. Unless SCP can be more easily purified than at present, and grown on substrates more economical than petroleum, it will not be a worldwide solution to protein shortages.

As discussed in earlier chapters, oils are expressed from a variety of plants, leaving a residue that is frequently used for livestock feed. Currently, most of this material is either contaminated or damaged by overheating, and is therefore useless for human consumption. However, with proper future development it is possible that such residues will be an important, high-protein food source. The residue remaining after the expression of oil from coconuts, for example, contains as much as 60% or more protein. In Mysore State in India and in Guatemala efforts are being made to convert such substances into acceptable foods containing 40% to 50% protein. Some species of oilseed plants contain substances that are harmful to humans, but in most cases it is possible to detoxify the materials. The extraction of a purified protein

Above *Plastic tunnels protect crops at the Horticultural Demonstration and Training Centre in Malta. The new Centre promotes the culture of vegetables, fruits, and flowers—with emphasis on production for export. (FAO photo)*

Left *Sterile fruit flies are laboratory bred and then introduced into areas plagued by the flies. The introduction of huge numbers of sterile flies into the population will result in the extinction of the pest. (FAO photo)*

powder from these residues might seem a feasible undertaking, but at present the unit cost of the protein thus extracted would be about five times the cost of producing the residue. This defeats the aim of a low-cost protein source. Again, there is also the problem of acceptability. The development of any relatively low-cost, high-protein foods will not meet human nutritional needs if people will not eat the substances. The Ehrlichs have stated that "the hungriest people are precisely those who recognize the fewest items as food" (Ehrlich and Ehrlich 1970). This suggests that solving the world's nutritional problems will not be as simple as providing enough quantity, although even that one aspect of the problem appears to be currently insoluble. Innovation in devising food sources will not succeed without acceptance.

Food from the Oceans   Phytoplankton harvesting has often been regarded as a potential means for meeting food needs. For various reasons, the assumption is a fallacy. Algae do not possess an especially great capacity for photosynthesis; but it is easier to disseminate the algae over vast areas of open ocean, to take advantage of surface sunlight, than to plant rows of seedlings. However, the cost of harvesting is then relatively high. There is also the problem of converting algae into popular foods. Perhaps the most important question centers on the wisdom of harvesting a plant substance that comprises the base of the food chain for most marine life.

The use of a greater variety of fish as food offers some possibilities, but not as many as proponents of the solution at times infer. Although more than 50 million tons of fish are caught every year, the consumption of fish by humans has actually decreased during the past decade. A meal can be made from fish, using "trash" fish that would not be acceptable as a market item. Such a meal is high in protein (80%) and additional nutrients, and can be incorporated into other foodstuffs to increase the overall nutritional value. There are some associated ecological problems that cannot be ignored, however. The harvesting of fish is now a relatively sophisticated process, with the use of radar and other technological aids to locate and capture fish in quantity. Although it was once assumed that the seas constituted a limitless resource, we now know how delicately balanced marine ecosystems are. It is entirely possible for modern fishermen to reduce fish populations to dangerously low levels. Presumably, gathering many heretofore unused fish varieties would increase the stress on marine ecosystems, since many of these commercially lesser fish are a food source for other, more desirable species. If every kind of fish is indiscriminately caught, then eventually many species will decline dramatically—not only as a result of the pressures of intensive fishing but due to a reduction in the food supplies of fish that are currently a regular commercial catch. It is worth mentioning, also, that water pollution and the destruction of estuary ecosystems threaten the populations of pelagic fish, as the young of many species mature in bays and estuaries.

Non-Plant Food Sources   Commercial rearing of terrestrial animals not customarily thought of as food has been attempted with minimal success.

New Animal Sources.   Capybaras, South American rodents that live along waterways, are large and tasty enough to have some food potential. Manatees and dugongs, entirely aquatic mammals that feed on vegetation of no direct value to humans, are other potential sources. The former live in freshwater environments of South America and Florida—browsing on water hyacinths and other plants; the latter are oceanic. Manatees seem a particularly appropriate possibility, as water hyacinths, upon which they feed, constitute a nuisance by clogging waterways. Attempts to breed manatees in Florida have met with no real success, however. In Africa, el.nds, a kind of large antelope, are an excellent potential meat animal, superior to cattle in several respects. They yield more protein per pound, and are adapted to grazing on marginal lands that are less efficiently exploited by cattle. Unfortunately, cattle are an integral thread in the cultural fabric of many African peoples. Convincing tribesmen to shift from cattle to antelope herding is certain to be a difficult if not impossible undertaking.

Insects as Food.   Insects could, in theory if not ever in practise, comprise an appreciable protein source. Grasshoppers, for example, are quite nutritious. The biomass of currently unexploited edible insect species is no doubt impressive, but it is not probable that even the most vigorous campaign could establish insectivorous habits among humans to any extent. Whatever the nutritional content, people are not liable to sit down and consume with relish plates of grasshoppers, termite larvae, lepidopterous grubs, and other such fare.

Marine Shellfish.   Among the better sources of limited or unexploited foods that could become more integral to diets in many areas are marine shellfish. Molluscs, such as mussels, can be seeded on heavy ropes that are dropped into offshore waters from floating stations. Later, the ropes are pulled up to deck when the mussels are mature. Such techniques provide the most feasible avenue for mariculture research, as open sea "fish farming" is not easily practicable; it is too difficult to control the environment. A disadvantage is that shellfish tend to be susceptible to pollution. For example, hepatitis can be contracted from eating oysters taken from water polluted with sewage.

Introducing New Foods To introduce successfully novel foods or to expand the importance of seldom eaten but nutritious plants or animals, there are four general parameters of consideration (Pirie 1967). First, research should be conducted privately, so that when popularization of the product begins there will be no doubts as to its merits. Second, those who are campaigning for the food should be known to eat it themselves, and the public support of influential individuals should be solicited. Third, there should be an abundant supply of the product available in advance of any publicity. Finally, economic factors are of particular significance, especially in persuading the citizens of developing countries to accept new foods. It makes little difference whether or not the foods are accepted if people cannot afford to buy them. So, in all cases, the cost of a new source of protein should be maintained at as low a level as possible.

A difficulty in introducing new food sources in the developing nations is that it is often less expensive and less trouble for governments to buy food from the developed countries. In New Guinea, fresh sweet potatoes cost more per calorie than imported wheat, and fresh fish is more expensive per gram of protein than canned fish. As long as these imbalances remain, governments of the developing countries are unlikely to invest in developing new foods. Even those politicians farsighted enough to realize that this cannot be a permanent situation are more often than not inclined to take the politically expedient course of providing the most at the lowest prices for the present—leaving future needs to be dealt with by someone else at some other time.

Whereas all potential food sources should be investigated, especially those high in protein, it would be unrealistic to assume that any of these could contribute significantly in meeting imminent food shortages. New food materials represent long-range possibilities; unfortunately, such materials offer no panacea for the current global nutrition crisis. Too much time is required for research, development, and to overcome or circumvent cultural eating habits. Rather than a solution, these foods represent a potential which may be better realized, hopefully, under a more stabilized balance between population and resources.

Suggested Readings

Ehrlich, Paul R., and Ehrlich, Anne H. *Population, Resources, Environment; Issues in Human Ecology.* San Francisco: W. H. Freeman and Company, Inc., 1970.

Pirie, W. N. "Orthodox and Unorthodox Methods of Meeting World Food Needs." *Scientific American* 216, no. 2 (1967): 27–35.

Wittwer, S. H. "Research and Technology on the United States Food Supply," in *Research for the World Food Crisis.* Edited by Daniel G. Aldrich, Jr. Washington, D.C.: The American Association for the Advancement of Science, 1970.

CHAPTER

22

FAMINE 1984?
NUTRITIONAL PROBLEMS
IN THE DEVELOPED AND
DEVELOPING COUNTRIES

An attitude of technological optimism flourished following World War II. People believed that hunger and malnutrition could be vanquished through the Green Revolution —leading to a world where everyone would have at least an adequate diet. Hybrid strains of wheat and other plant crops produced much greater yields than the nonengineered varieties. Hybrids could also be developed for specific environments. Why, then, instead of the realization of so blissful a dream, are we confronted with the bitter reality that our cornucopia is probably emptier than it has ever been?

Overpopulation There is no single answer; but a major component is human overpopulation, a threat to our lives in a variety of ways. On July 24, 1969, when Neil Armstrong took his first steps on the moon's Sea of Tranquility, this coup of science and technology was applauded as an indication of our potentially limitless capacity for delving into the mysteries of the physical universe. But consider this fact: during the total eight-day mission of Apollo II, the world's population increased by one and a half million, and an estimated 100,000 people starved to death. We are able to explore the fringes of outer space, but we have made insignificant progress in solving the critical problems of all nations, including the U. S. "No other problem facing mankind today is so crucial as the food and population problems," commented Dr. W. M. Myers, a scientist associated with the Rockefeller Foundation, in a 1968 symposium of the American Association for the Advancement of Science. "These two problems are the roots of many others—domestic and international unrest, the widening gap between the 'haves' and the 'have nots,' the lags in economic development of the lesser developed countries and of underprivileged segments of this country. Unless these problems of food supplies and population growth can be solved, all other efforts to build a better world will come to naught" (Myers, 1970). We cannot ignore the fact that an increasing percentage of humanity is dependent on the grain exports of a few countries (Wortman, 1976).

Dr. Myers then went on to express his belief that these problems were unquestionably soluble, and that agricultural production could, within only a few years, close the nutrition-population gap. His was a perspective of eight years ago, and it is understandable why he remained a devotee of technological omnipotence in food production. However, our contemporary perspective does not offer a similar hope to an objective student of the facts. In previous chapters we have described the flaws inherent in depending too completely on hybrid strains of crops, and we have examined other reasons

FAMINE 1984?

why the Green Revolution has not produced the results that it seemed to promise. In the context of a mushrooming global population, of which millions already starve to death annually, we do not find any evidence that portends a happy ending for the drama of agricultural resource-population imbalances. Although the real severity of this crisis has made its impact only within the last several years, a number of informed individuals have long foreseen existing food scarcities as an inevitable outcome of population growth and technological misdirection. The British writer C. P. Snow predicted in 1968 that the beginnings of worldwide famine would occur between 1975 and 1980 (Snow 1968).

The calamities inherent in continuous population growth were recognized, in generalized fashion, centuries ago. Thomas Malthus warned in 1798 that increasing population would culminate in famine and starvation. From a most basic perspective, if the rate of population increase is greater than the death rate, an expanding number of people will occupy the earth at any one time. Malthus correctly assumed that agricultural resources, even at peak productivity, could be augmented only on an arithmetical basis—that is, a simple expansion of the original quantity (2, 4, 6, 8, 10, etc.)—whereas the human population increases geometrically, or exponentially (2, 4, 8, 16, 32, etc.). Although many factors influence the reality, which is not as precise as a mathematical formula, the basic concept holds true. Accordingly, there is a point in population growth when the population exceeds its resources. Malthus' answer for the catastrophes he foresaw was to work toward "the greatest good for the greatest number." Unfortunately, this solution requires the maximization of two variables, which cannot be done. "The greatest good for a strictly limited number" is the only workable goal of the present. "Maximum welfare," advised Arnold Toynbee in *Man and Hunger* (1963), "not maximum population, is our human objective."

Why has the world's human population become a seemingly insoluble problem at almost the same time that the Green Revolution offered the expectation of relieving humankind from hunger? The nature of exponential increase, coupled with improved medical care aiding longevity and sharply reducing infant and child mortality, are the principal explanations. It is estimated that 10,000 years ago there were approximately 5 million people inhabiting the earth—about a third as many as currently dwell in Mexico City. At that time, the "Agricultural Revolution" occurred in several unrelated areas of the world—the Near East, Asia, Mesoamerica, and perhaps in other regions—giving human beings their first prolonged experience with relative freedom from searching for food, and a concomitant advantage of free time. A sedentary agricultural lifestyle eased some of the restraints on population, and the number of the human species soared to a level of 133 million by the time of Christ, a 26× increase. In the 2000 years since, the world population has grown to the present figure of nearly 4 billion. It has been estimated that a million years passed from the time that human beings numbered slightly more than 1 million, until the Agricultural Revolution when there were 5 million. Currently, the population of 4 billion is expected to double in 35 years, unless the more probable outcome of famine, with the corollaries of disease and starvation, reaps a grim harvest.

Even the simple geometric increase predicted by Malthus would eventually result in staggering numbers, but in fact the actual rate of increase is rising. According to the United Nations Department of Economics and Social Affairs, the world average rate is somewhat less than 3%; but in some countries, such as Brazil, it is more, and in Western industrialized nations it is only about 1% or less. Even at 1%, we believe the rate is too high. One's first impression of that figure may be that it is quite low, an application of the figure suggests otherwise. For, if the population of the United States were to continue to increase at 1%, it would double during the next 70 years! Even at a true rate of zero population growth (ZPG) in 1976 there would be a 70 year lag to 2035 before birth and death rates equalized. During that time another 70 million people would have been added to the present population of 220 million. The only viable answer for the crises arising from overpopulation is global ZPG; any rate of increase, however small, must eventually lead to too many people.

In the past, the human population has tended to stabilize after each "revolution" which gave impetus to population growth. However, there seems no such hope now. The rate of increase is too great, and the numbers too vast to suppose that any inherent limitations will manifest themselves, beyond those of catastrophe.

It is true that the logistics problem of food distribution are partially responsible for the fact that millions of people die from starvation yearly. However, too many people and too little food constitute the principal cause of worldwide hunger, and every year hence will see this problem exacerbated. At present, agricultural production can barely keep up with global needs. If there is not enough food for 4 billion, how will we provide the quantity needed by twice that number, or more? In *Alice In Wonderland*, Alice finds that she must run faster just to stay in the same place. With our food-population predicament, we may find that even by "running faster"—producing more food—we will not even stay in the same place, but will fall behind.

Environmental Feedback   To comprehend better the failure of agricultural advances to feed the world's proliferating masses, let us consider not only overpopulation, but also the interactions among factors such as land use, water use, the failure to recognize environmental limitations, and community and global economics. Doubling the population *more* than doubles the feedback on these factors. Agricultural practices which ignore environmental constraints, even if the practices are highly productive for a few years, eventually culminate in reduced or zero fertility; thus decreasing food production. Irrigation in arid regions, for example, may cause an accumulation of salts in the soil, resulting in land useless for agriculture. This has already happened in West Pakistan, where technology was not adequate to avert the problem. More than 5 million acres of land, or 18% of the total cultivated in that country, is out of production for this reason. Salts are pulled up to the surface, through capillary action, when the land is carelessly or inadequately irrigated, and eventually not even native vegetation will grow.

A major consideration in the Green Revolution is that a large percentage of the earth's land surface is not adaptable to high-yield agriculture, even under optimal circumstances. Making "gardens" of the desert has been successful in a few areas, such as Israel; but in the meantime climatological changes have been making deserts of gardens, or at least of areas that once were able to support restricted populations.

From Garden to Desert   The much-publicized Sahel region in Africa is the best contemporary example of a substantial area rapidly becoming worthless to humans. The Sahel lies just below the Sahara desert, and has long been a relatively harsh environ, precariously dry and requiring an adaptive ingenuity from the nomadic tribes sustaining themselves there. The annual rainfall, however limited, allowed for a sparse growth of natural vegetation, enough to feed limited herds of cattle, camel, and goats, and therefore a restricted population of small tribes. Several years ago, the annual rains began to fail. Weather in the Sahel has been governed by the movements of bodies of cold polar air. During spring and early summer, the cold air characteristically receded to the north, drawing up warmer air masses, followed by moist air belts which brought monsoon rains. Then, in the fall the cooler air would flow south-

Above *Sheep, one of the first domesticates, create tremendous grazing pressures on natural environments. The population densities are frequently high and the animals' destructive grazing habits result in the uprooting of native vegetation and subsequent erosion. (FAO photo)*

Left *Dry, cracked land along the Awash River, Ethiopia. (FAO photo)*

ward once again, presaging the dry season. This pattern, although not providing abundance, brought enough moisture to the Sahel for the scattered populations of people and domestic animals to eke a living from the arid land.

Apparently the cooler air masses are no longer moving northward, and the Sahel's dry season has been extended to a year-round condition. In Boutilimit, Mauritania, which is far into the Sahel, rainfall decreased from a previous average of 200 ml (milliliters) during July through September, to 41.6 ml in 1973. Reductions in other parts of the Sahel were equally drastic. Shifting, moving sand dunes, dangerous portenders of the starkest kind of desert, have crept from the Sahara throughout the Sahel, even invading Kenya. The Sahara has claimed a boundary 60 miles south of the pre-drought limit. Bleaching bones of animals lie amidst the abandoned ruins of villages, all disappearing under the parched sand that has advanced to expand the world's largest desert, an area greater than that of the United States. Once this same area provided grain for the Roman Empire. Deforestation, poor agricultural practices, and climatological change have reduced it to its present state.

Whether or not the Sahel drought is a temporary condition or the result of long-range changes cannot yet be proven one way or another. A cooling trend has been affecting us since the 1940's. Many, but not all, meteorologists believe that it is the result of atmospheric pollution—such as smog and dust particles—causing a greater reflection of the sun's warming rays. Whatever the cause, an overall global temperature reduction of only a few degrees could augur future agricultural disasters for many nations in northern latitudes. The last ice age resulted from a drop of only 5° to 7°C.

Undeveloped Land Potential   Other vast and, as yet, undeveloped regions, such as the Amazon and the Congo Basins, have been greatly overrated in terms of potential productivity. With complete development each of these areas could feed another 200 million people, under ideal conditions; but it will take 50 years to accomplish. In the meantime, at present rates of increase, there will be 5 *billion* more people to be fed (Borgstrom 1973). There is no way to close the discrepancy between arithmetical and exponential increases, however great the effort!

Most of the truly fertile areas of the world are already under cultivation. Deserts provide minimal hope, due to the problems inherent in ever-expanding irrigation. Tropical jungles do not have the potential productivity that had been expected by agricultural experts. On first thought, it might seem that a rain forest, which has thousands of species of plants, would yield extraordinarily fertile ground when cleared. Unfortunately, the reality is quite the opposite. However lush the forest may seem, about 80% of the nutrients in a jungle ecosystem lie above the ground. The soil beneath the forest, instead of being rich in nutrients, is of much poorer quality than a typical fertile farm soil and does not readily support high-yield crops.

Another serious effect of developing tropical lands is the tendency for the soils to laterize—being transformed into a rocklike material called laterite (from the Latin *later*, meaning brick). When the jungle has been cut away or burned off, the exposed soil is rapidly leached of its nutrient minerals. Laterization then results from the exposure to sun and oxygen which produces complex chemical interactions. Once effected, this ruins the soil for agricultural purposes. The temples of Angkor Wat, built by the Khmers between 800 to 1,000 years ago, are composed of sandstone and lateritic stone. Laterization of the soil has been proposed as the principal reason for the disappearance of the Khmer civilization.

The primitive "slash-and-burn" agriculture of tropical regions is conservative ecologically. A small plot is cleared of low growth and burned, a technique that usually leaves larger trees standing. After several years of use, the soil's productivity has been reduced and the plot is abandoned. Complete laterization will not have taken place in this interval, so the forest rapidly reclaims the land. This reclamation prevents further conversion of the soil into rock, and aids in the disintegration of laterite that may already have formed.

As with a great many "primitive" agricultural practices, the slash-and-burn method is not too out of tune with environmental requirements, despite the relatively low yield when compared with the harvests of modern mechanized agriculture. Of course slash-and-burn is an operable system only in areas of light population density. Otherwise, native vegetation would never have a chance to resume growth. In many parts of Latin America, there are endless denuded hills in areas where there are so many people that land cannot be spared to renew itself.

The Amazon Basin, currently the focus of an intensive development campaign by the Brazilian government, has been considered by some to be an answer to South America's food scarcity. However, as we have already pointed out, even under the best of conditions the basin would produce a drop in the bucket of world needs, and a remarkably diverse wilderness area would be lost forever. It is extremely doubtful if the land will live up to the developers' hopes, for several reasons: the poor quality of the soil; laterization; and flooding of the Amazon. Laterization is almost certain to effect development of the Amazon Basin far more than the Brazilian government apparently recognizes or accepts. There is a continuing campaign to clear the jungle. Iata, in the Basin, represents a government-planned attempt to establish a farming community in the jungle. It failed dismally. Within five years "the cleared fields became virtually pavements of rock" (Ehrlich and Ehrlich, 1970). Also, the spread of mechanized agricultural practices in that region will surely cause environmental havoc, leading eventually to a deforested land of little or no agricultural value. Mechanization will facilitate rapid exploitation, and will cause greater erosion of the soil's limited nutrients.

Annual flooding of an extensive area may also be expected, because the river's natural dike—the jungle—will have been removed. In 1974, the river overflowed as never before, flooding villages and drowning an uncounted number of people and livestock. At that time, the Brazilian Nature Conservancy recognized the potential for full-scale disaster in the future and called upon the Brazilian government to desist from its reckless frontier philosophy. Nevertheless, the official attitude in Brazil remains behind the proponents of wilderness development.

The term "overpopulation" is a relative concept, referring not just to the number of people per unit of land but to the interrelationship between population, resources, and environment; so the quantity of "undeveloped" land remaining is meaningless agriculturally except in the context of what can be expected from that land in terms of food production. That there are millions of acres in the Amazon Basin that have not yet been exploited for agriculture signifies, as we have seen, very little in relation to combating global food shortages or even in preventing South American nutritional inadequacies. In most cases, human beings would be wiser to avoid the cultivation of marginally productive lands altogether. Ecological diversity is desirable for a healthy global ecology, from which we cannot exclude ourselves; and the farming of these areas will not solve the food-population crisis.

Effects of Malnutrition Since much of the world's agricultural productive potential has already been realized, one cannot be too optimistic about the probability of ending famine in the next several decades. About half of the world's people are suffering from malnutrition (receiving enough calories to live but not sufficient critical food constituents, such as protein) or starvation (not enough calories or essential nutrients). Individuals in this category are perpetually tired and highly vulnerable to disease, which kills most starving victims first.

Mexico: A Developing Country   Our neighboring country, Mexico, serves as an illustration of some myths about the advantages of "development" (Borgstrom 1973). Mexico City, with a population of nearly 15 million, is the largest city in the Americas, and seethes with the bustle of industry, construction, and expanding wealth. Boulevards are wide, flowing with late-model autos, and lined with attractive shops.

This surface luster covers a variety of less pleasant realities. Mexico has been applauded for an agricultural yield surpassing the rate of its population growth, but the average harvest per acre has not increased; only the amount of land under cultivation has increased. In a population of somewhat more than 50 million, half still live chiefly within a local economy. Twelve to 15 million people earn ten dollars per month or less, and three out of five families earn 80 dollars per month. One half of the families in Mexico City live in a single room, and another quarter have two rooms. Mexico's affluence is shared by a relative few. The country actually comprises two economies, one akin to more affluent Italy, the other similar to underdeveloped Asia.

Despite the appearance of widespread economic growth, more than three-fifths of Mexico's population is still malnourished and undernourished. The glowing pictures of Mexico's economy are not depicted by the poor. Thomas Malthus observed that the deprived and poor are the first to bear the burden of shrinking natural resources, yet are never the ones who project the impression of reality; if they could, we would have a markedly different view.

Among Latin American countries, Mexico is relatively affluent, but still more than half of the population is nutritionally deprived. The illusion of well-being that masks hunger is in itself a problem in defeating hunger. The poorest of Mexico's poor eat little more than corn, primarily in the form of tortillas. This is an incomplete protein diet. Can Mexico hope to achieve true affluence or even modest well-being for everyone? At its present growth rate, the current population of 50 million will be 200 million by the year 2000. Even if all virgin land in Mexico were cultivated, it is doubtful if food needs could be satisfied for so great a number.

Many have believed that the way to vanquish poverty is to increase industrialization. However, we must now face the consideration that industry is no longer a necessarily desirable goal for developing countries. Further, successful industrialization cannot be accomplished without an improvement in agricultural efficiency. In most developing countries, 60 to 80% of the people are engaged in agriculture; yet they produce much less than is needed for their own use. A country at that level cannot industrialize before it has solved the problem of providing enough food for the people's needs. Industrialization as an economic panacea is an anachronistic approach (Ehrlich and Harriman 1971).

Africa

Africa is presently a prime area of malnutrition, and the future looks bleak, in spite of a relatively low population density. Within 11,685,000 square miles, there are less than 400 million people. Yet, due to political conflicts and the low productivity of much of the land, Africa is already over-populated. The second-largest continent, it stretches across the equator, which serves nearly as a median dividing line. The terrain varies greatly—from the immense Sahara to the thick, humid equatorial jungles of the Congo, to the plains of Kenya. Accordingly, there is no single formula for the agricultural-nutritional problems besetting Africa's diverse populace.

The Sahel reflects a unique crisis resulting from weather change and overdevelopment, but that tragedy is not indicative of the factors contributing to most of Africa's nutritional deprivation. Dahomey, the small country between Nigeria and Togo, lapped on its southern shore by the Gulf of Guinea, is more representative of the general problems of the non-industrialized African nations, although each country has a specific set of varying conditions (May 1968).

Dahomey: An African Nation. Somewhat smaller than Pennsylvania, Dahomey is populated by less than 3 million people. Its economy is one of the most purely agricultural in Africa. Nearly all of the 3 million native inhabitants, except for very young children, are in some way engaged in agricultural pursuits. About 17% of the land is considered arable, and most of that is planted in oil palms. The oil from these is Dahomey's only significant cash crop, and the only export of any impact on the economy. The palms are well adapted to the sandy coastal soil—growing where most other commercial crops would not easily gain foothold.

Corn is the primary food crop of the south, and millet predominates in the north. Other crops, grown almost entirely for local consumption, for individual families or at the village level, include yams, manioc, sweet potatoes, taro, legumes, other vegetables, and a variety of tropical fruits, such as citrus, mango, and papaya, all important dietary constituents. Most of these foods have a negligible protein

Above *Farmers at IRAT
(Institute Recherche
Agronomique Tropical) in
Logozohe, Dahomey, store
sunflowers to dry and use as
cattle fodder. (FAO photo)*

Left *Mother and children at a
clinic in southern Dahomey.
(FAO photo)*

content, but they are what most Dahomians depend on for sustenance. There are some cattle, chickens, and other domestic animals, but not enough to affect economically the whole population.

Dahomey's malnutrition level is rising, due to sociological factors as well as the problems of increasing food production. Most farmers formerly practiced shifting agriculture, in which one field is allowed to fallow and renew its productive capacity while another is being used. People lived in small villages and were allotted their share of tillable land by a headman. Property was communally owned, and food resources were shared evenly among villagers. In the past several decades, there has been a shift to private ownership of plots, a change resulting in less efficient use of agricultural acreage and less equitable distribution of resources. This alteration in the traditional rural pattern has caused increasing migration to cities, where most of the formerly rural inhabitants live in slums, and are thereby condemned to malnutrition and poverty. With the continuation of this trend, the inadequacy, both in quantity and distribution, of food resources is almost certain to become more severe.

On the surface, Mexico and Dahomey are in quite different circumstances; yet in both cases a majority of the population is not adequately fed. In the United States, for example, existing malnutrition is frequently a result of lack of knowledge of nutrition; not the result of the inability to obtain a satisfactory diet. Even the poorest of citizens in the U. S. have some recourse, in the form of food stamps and other welfare. In contrast, in most developing nations there is little or no government assistance for the hungry millions.

Every nation's food problems vary in some respects from other nations; but with scarcity of food, the results are the same. A starving child in Biafra feels as much pain as a starving old woman in India. The face of hunger, grotesque and enervated, is one face behind the masks of race, nationality, ethnic group, religion, or any of the other tenuous barriers dividing humankind.

1984?

Is world famine in 1984, or sooner, an inevitable culmination of population growth exceeding resources? If the growth continues, the answer may be "yes." There are already famines scourging the struggling peoples in parts of Asia, Africa, and other scattered areas of the world. At this time, there is no foreseeable remedy except a reduction of human numbers coupled with increased agricultural efficiency and, hopefully, a shift toward staple crops higher in protein than corn, wheat, and rice.

We cannot ignore the question of quality. People need more than calories. They need a minimal diet of critical amino acids, vitamins, and other nutrients. Attempting to arrive at an average based on variations in size, metabolism, and so forth, the United Nations Food and Agricultural Organization has placed the daily individual caloric need at 2354. In reality, some people will need considerably more to remain in good health, and others less. On one hand, meeting the minimal caloric intake is imperative, but on the other it is a hollow goal unless the calories are accompanied by the equally necessary minimum of basic nutrients. This is true for all of humankind. A person cannot subsist in reasonable health on bowls of rice alone. Many people have little more than that to eat every day, but their basic requirements are the same as ours.

Perhaps the most hopeful possibility for solving the problem of energy-conservation vs. protein-content is to develop grains with higher percentages of more complete proteins. In 1963 a high-lysine gene, labeled Opaque-2, was discovered in corn, through research at Purdue University. Scientists associated with the Rockefeller Agricultural Research Program in Colombia developed varieties of corn with increased lysine, and thus a more balanced protein content containing all of the essential amino acids. These were released for general use in 1969.

Even with break-throughs of this nature, may we reasonably hope to expand food production enough to catch up with and overtake population increase? From the perspective of the contemporary global food situation, the answer is, "probably not." To achieve this end, not only would more food have to be grown, but distribution methods would have to be greatly improved; food preservation techniques expanded, especially in tropical countries; and the quality of crops and livestock changed through hybridization to yield more protein for the energy consumed in development. Also, a crucial aspect of dealing with the hunger crisis is that cultural attitudes and traditions frequently stand in the way of solutions. Corn, for example, is so completely woven into the cultural fabric of most Latin American countries that persuading rural farmers to use their land for other crops higher in protein is often difficult if not impossible. Beans, which contain more protein than corn, are a popular diet as well, but traditionally are of secondary importance to corn. Even the protein in beans needs to be complemented with additional amino acids. These may be supplied through milk products or small amounts of meat. In Latin America, the following ingredients have been added to corn meal: vitamin A, 3% fish meal, 3% egg powder, 3% food yeast, 5% skim milk, 8% soy bean flour, and 8% cotton seed flour. This combination has been moderately successful in combatting protein deficiencies, although the added cost and culturally inherent reservations about new foods have prevented large scale success.

One temporary means suggested for aiding countries currently experiencing devastating food shortages is to establish a world food bank, which would be supplied by the United States and other agriculturally sufficient countries. There would still be problems in shipping the surpluses; and there would have to be methods for storing and distributing the food. Most of the countries hardest hit by food scarcity are the least equipped for effective distribution of donations. In the Sahel, the transportation of truckloads of grain into the interior, where most roads are unpaved and extremely rugged, has met with only limited success. However, a world food bank is a better immediate solution than none at all.

"A world food bank," suggests Roger Revelle (1974), "should have several components—stores of wheat and other cereals and of soybeans and other legumes; stores of fertilizers to enable crop production to be quickly expanded; reserves of land which can be put under the plow in emergencies; a bank of information and technology which can be used to increase crop yields; and stores of crop genes to enable seeds of new varieties to be quickly multiplied when the old varieties are stricken by pests or plant disease."

These are laudable aims. However: "between the Idea and the Reality, lies the Shadow" (T. S. Eliot). Much of the global population is already starving, and presumably most land which can be immediately "put under the plow" already has been. There is a worldwide shortage of fertilizers, so it would be difficult to accumulate such stores. A world food bank should be established, but it would probably be quite limited in what it could accomplish to ease hunger. The increase in world grain production has averaged 30 million tons annually, yet food reserves have dwindled (McLaughlin 1974). U. S. federal spending on food programs has increased from 1.6 billion dollars in fiscal 1970 to 5.1 billion in fiscal 1974, and still more people are dying of starvation (New York Times 1974).

There is famine today, and there probably will be in 1984. Greater agricultural production and foods containing higher percentages of complete proteins and other nutrients are a boon to humankind, but the potential of these foods cannot be realized until the rate of increase of worldwide population has been reduced to zero. At this point, any further population expansion carries with it the seeds of disaster, undermining our attempts to sustain an agricultural abundance.

Suggested Readings

Aldrich, Daniel G., Jr., ed. *Research for the World Food Crisis.* Washington, D.C.: American Association for the Advancement of Science, 1970.

Borgstom, Georg. *Focal Points.* New York: MacMillan Publishing Co., 1973.

Brown, Lester R. *Seeds of Change.* New York: Praeger Publishers, 1970.

Ehrlich, Paul R., and Ehrlich, Anne H. *Population, Resources, Environment; Issues in Human Ecology.* San Francisco: W. H. Freeman and Co., 1970.

Ehrlich, Paul R., and Harriman, Richard L. *How to Be a Survivor.* New York: Ballantine Books, 1971.

Hardin, Clifford M., ed. *Overcoming World Hunger.* Englewood Cliffs, N.J.: Prentice-Hall, Inc., 1969.

May, Jacques M. *The Ecology of Malnutrition in the French Speaking Countries of West Africa and Madagascar.* New York: Hafner Publishing Co., 1968.

McLaughlin, Martin M. "Feeding the Unfed." *Commonweal* 100 (1974): 376–399.

McCarthy, Terence. "Feast a Famine: The Choices for Mankind." *Ramparts* 13 (1974): 29–32, 59–60.

The New York *Times* (editorial). "Hunger in America." July 8, 1974.

Revelle, Roger. "Food and Population." *Scientific American* 231 (1974): 160–170.

Revelle, Roger. "Will There Be Enough Food?" *Science* 184 (1974): 1135.

Scientific American 235 (1976). (The entire September 1976 issue is devoted to aspects of food and agriculture.)

Simon, A. "Food Crisis." *Commonweal* 100 (1974): 374–375.

Snow, C. P. Unpublished paper presented at John Findley Green Lecture, Westminster College, Fulton, Missouri, 1968.

Wortman, Sterling. "Food and Agriculture." *Scientific American* 235 (1976): 31–39.

INDEX

coconut fibers, 133
Cocos nucifera, 153–155
Codex Mendoza, 176
coevolution, 53, 60
Coffea, 207–209
coffee, 207–209
Coleoptera, 53
collenchyma, 29
Colocasia esculenta, 167
Columbus, 186
Conchocelis, 317
Congo, 335
Consultative Group on International
    Agricultural Research, 308
consumer attitudes, 295
copra, 155
coprolites, 66
corm, 31
corn, 89, 107–113, 265, 266, 268
cotton, 29, 125–127
Coxcatlan, 90, 178
crabgrasses, 121
Cretaceous Period, 51
Cronica del Peru, 162
Crystallic Period, 85
Cucumis, 161
cultigens, earliest, 72
cultivars, 25
curanderas, 256
Curcuma longa, 199
Cyanophyta, 317
cycads, 49
cyclones, 290
cytology, 23
cytoplasm, 27
cytoplasmic sterility factor, 111

Dahomey, 339–341
Dahomey, South, 235
Darwin, Charles, 46
date, 153
date, Chinese, 153
Datura, 245
Daucus carota, 167
de Candolle, 155
de Leon, Pedro, 162
De Materia Medica, 257

de Mendoza, Antonio, 176
dendrochronology, 134
Densmore, Frances, 254
diatomaceous earth, 316
diatoms, 316
Diaz, Porfirio, 93
Dick, Herbert, 89
diet, alternatives, 273
    changing human, 262, 263
    Mesoamerican, 265–267
    Middle Eastern, 267
    upgrading of, 271
dietary innovation, 321–328, 323, 324
dimethyl tryptamine, 247
Dioscorea, 165
Dioscorides, 257
diploid, 33
distilled spirits, 215, 217
DMT, 247
Doctrine of Signatures, 258
dogbane, 255
Drew-Baker, K. M., 317
drug use among primitive peoples, 247
dugong, 326

Ehrlich, Paul, 276, 289
    Paul and Anne, 325
Einkorn wheat, 104
Eliot, T. S., 343
energy conservation, 16, 17, 298–300
epinephrine, 245
Esbats, 254
Escherichia coli, 316
"Essays on The Principle of
    Population," 290
essential oils, 196, 201–203
    various minor types, 203
ethnobotany, 16, 222–228
    field work in, 222–228
Eugenia caryophyllus, 198
eukaryotic cell, 312
Euphrates Valley, 104, 258
eutrophication, 317
evolution, 46
exine, 79
Extensionistic Period, 85

family, 26
famine, 341

FAO, 308, 319, 342
farinha, 166
fertile crescent, 76
fibers, 124, 125
Ficus carica, 158, 159
fig, 158, 159
    Smyrna, 158
finches, Darwin's, 12
fish meal, 325
fish, smoked, 269
flax, 127–128
flower, 31
Food and Agricultural Organization
    (FAO), 308, 319, 342
food bank, 343
food chains, 16, 17
food surpluses, 12, 73
Ford Foundation, 283
forest clearance, 79–81
forestry, 135–137
Frageria, 148
Frankel, O. H., 308
Frederick 11, 258
fruits, 145–147

Gaines wheat hybrid, 105, 280
Galápagos Islands, 12, 98
gall flower, 159
gametes, 31
Ganges River, 290
Gelidium, 313
genes, 47
genetic conservation, 307, 308
genetic drift, 26
genetic variability, loss of, 304, 305
genetic variation, sources of, 309
genome, 37
genus, 23, 25, 26
Gerard, 162
gin, 217
ginger, 119
ginning, 124
ginseng, 259
gliadin, 105
glutenin, 105
Glycine max, 178–180
    usurriensis, 178
Goodyear, Charles, 191

Gossypium, 125
grafting, 43
Green Revolution, 279–291
    results of, 283–285
Guatemala, 107
gum arabic, 257

half-life, atomic, 67
hallucinogens, 244–254
    and witchcraft, 253, 254
haploid, 33
"hard seed", 177
Hardin, Garrett, 15
Hardin, Lowell, 283
harmaline, 244
harmine, 244
Harner, Michael, 253
healing, 254–256
healing and power, 256, 257
hemicellulose, 27
hemp, 131–133
    Manila, 133
henbane, 245, 253
henequen, 129
Herbal, 162
Herodotus, 257
Hevea brasiliensis 191–193
Hippocrates, 201
History of Plants, 22
Hoabinhians, 84
Homo erectus, 67
    habilis, 67
    neanderthalensis, 67
    sapiens, 68
hops, 213
horseradish, 169, 199
Humulus lupulus, 213
hunger, economics of, 285, 286
Huxley, Aldous, 245
hybrid, 47
hybrid grains, 285
hybridization, 25, 47, 108
Hymenoptera, 53
Hyoscyamus, 245
hypogynous, 51

Ilex paraguariensis, 212
INCAP, 271

Incaparina, 271, 285
Incas, 148
India, 283
indole acetic acid, 44
Indus Valley, 104
innovative foods, introduction of, 327
Inquisition, 244, 253
insects, 53, 55
  as food, 326
Institute of Nutrition of Central
  America and Panama, 271
International Code of Botanical
  Nomenclature, 23
International Rice Research Institute
  (IRRI), 115, 280
Ipomoea batatas, 163
  trifida, 163
Iridaea, 319
IRRI, 115, 280
isolating mechanisms, 49
isolation (genetic), 47
isoleucine, 263
isotope, 67

Jarmo, 76, 104, 176
jebong system, 192, 193
Jivaro Indians, 249
Job's Tears, 121
Jurassic Period, 49
jute, 133

kapok tree, 133
Kensinger, Kenneth, 249
Kew Gardens, 192
Khmers, 17
Kihara Institute for Biological Research,
  83
Kikuyu farmers, 270
kombu, 317
kraft process, 137
kwashiorkor, 172, 267

lamella, 29
Laminaria, 315, 316, 317, 319
Late Classic Period, 95
leaf, 31
legumes, 172–183, 265

Leguminosae, 172–183
Lens culinaris, 175–176
lentils, 175–176
Leopold, Aldo Starker, 300
Lepidoptera, 53
leucine, 263
licorice root, 258
Lignic Period, 85
lignin, 27, 134
Linnaeus, Carolus, 23
Linne, Carl von, 23
Linum usitatissimum, 127
Lithic Period, 85
loco weed, 245
Lophophora williamsi, 244
LSD, 244, 245
Lummi Indians, 318
lycanthropy, 254
Lycopersicon esulentum, 149
lycopods, 49
lysergic acid diethylamide, 244, 245
lysine, 263, 274

MacIntosh, Charles, 191
MacNeish, Richard, 87
Macrocystis, 315
macroenvironment, 69
mágicos, 256
maguey, 300
maize, 89
malnutrition, 267, 268
  effects of, 269, 270, 337, 338
Malthus, Thomas, 290, 331, 332
Man And Hunger, 331
Manáos, 192
manatee, 326
mandazi, 268
Mandragora, 245
mandrake, 245
Mangelsdorf, Paul, 72, 87, 107
Mangifera indica, 152, 153
mango, 152, 153
manido, 254
Manihot esculenta, 165–167
manioc, 165–167
Maoris, 163, 258
maple sugar, 189–191
margarine, 179

pedicel, 31
pellagra, 265
pepper, 197
perfumes, 196, 201
pericarp, 105
Period of Conflicting Empires, 86
Persea americana, 147
Peru, 71
pesticides, 15, 299
peyote plant, 244
Phaeophyta, 315
pharmacology, 257
Pharmacopeia, 258
Phaseolus, 176–178
    coccineus, 177
    lunatus, 177
    vulgaris, 177, 268
phenylalanine, 263
Philippine Rice and Corn
    Administration, 284
phloem, 29
Phoenix dactylifera, 153
photosynthesis, 29
Phytophthora infestans, 162
phytoplankton, 325
pineapples, 133, 148
Piper nigrum, 197
Pisum sativum, 175
plant domestication, 68, 69–71, 77
Pleistocene Epoch, 15, 262
Pliny the Elder, 201
plywood, 137
Pogostemon cablin, 202
pollen records, 79–81, 89
pollination, 53, 55, 59
polymer, 27
polyploidy, 33
pomades, 202
population explosion, 290, 330–333
Porphyra, 317
potato, Irish, 162, 163
    sweet, 163–165
potato blight, 162
Pouyata, Jean Francois, 157
primary center, 309
protein, 263, 264, 265, 275, 342
    compensatory plant, 265
    plant vs. animal, 275
    quantitative requirements, 264

Prunus, 160
pseudocopulation, 55
psilocin, 244
Psilocybe mexicana, 244, 250
psilocybin, 244, 245
psychedelic experience, 245
Psychotria viridis, 247
Puccinea graminis, 104
Pure Food and Drug Administration,
    258
Pyrrophyta, 317
Pyrus, 161

Quintana Roo, 94

radishes, 169
raised fields, 95
ramie, 128
Ranales, 51
Raphanus sativus, 169
receptacle, 31
recombination (genetic), 47
reduction, 31
reproduction, 31, 41
resellers, 231, 232
retting, 124
Revelle, Roger, 343
Rhizobium, 173
rhizome, 31, 43
Rhodophyta, 313
rice, 83, 113–117, 268
    modern cultivation of, 114
    paddy, 114
    various uses of, 115, 117
"rice paper", 117
Río Bec, 94
rippling (of flax), 128
Rockefeller Foundation, 280, 308, 330
root, 31, 255
root crops, 161–169
Rorippa armoracia, 169
rose oil, 202
rubber, 191–194
    extraction of, 192, 193
    various minor sources, 193
    vulcanization of, 191
rum, 217
rutabaga, 169
rye, 119

tobacco, 140–142
tofu, 274
Tolypothrix, 316
tomatoes, 149
tools, invention of, 67
topsoil, 15
topsoil conservation, 95, 300
Toynbee, Arnold, 331
Triassic Period, 49
Tripsacum, 89, 107
Triticale, 107
Triticum aestivum, 104
    monococcum, 104
truckers, 231
tryptophan, 263
turmeric, 199
turnip, 169
tuber, 31, 43

United States Department of
    Agriculture (USDA), 307
USDA, 307

valin, 263
vanilla, 198
Vanilla planifolia, 198
vanillin, 198
variety (taxonomic), 23, 25
vegetative propagation, 41
vermouth, 215
"Victoria type" oat rust, 304
Vitis vinifera, 215

wasp, chalcid, 159
weaving, 124, 125
wheat, 77, 104–107
    milling of, 105
wheat rust fungus, 104
whiskey, 217
Wickham, Sir Henry, 191
"wild rice", 121
wild strains, importance of, 305
wine, 214, 215
witchcraft, 244
women as leading figures in peasant
    market economy, 235
wood, 133–135
Woolman, John, 125
wort, 213

Xanthosoma sagittifolium, 167
Xochimilco, 93
xylem, 29, 133

yams, 165, 167
yeast, 107

Zea mays, 89
zein, 172
zero population growth (ZPG), 332
Zingiber officinale, 199
Zizyphus jujube, 153
ZPG, 332
zygote, 33